T0245393

Bioinformatics and Computational Biology in Drug Discovery and Development

Computational biology drives discovery through its use of high-throughput informatics approaches. This book provides a road map of the current drug development process, and how computational biology approaches play a critical role across the entire drug discovery pipeline.

Through the use of previously unpublished, real-life case studies, the impact of a range of computational approaches are discussed at various phases of the pipeline. Additionally, a focus section provides innovative visualization approaches, from both the drug discovery process as well as from other fields that utilize large data sets, recognizing the increasing use of such technology.

Serving the needs of early career and more experienced scientists, this up-to-date reference provides an essential introduction to the process and background of drug discovery, highlighting how computational researchers can contribute to that pipeline.

William T. Loging is Associate Professor of Genetics and Genomic Sciences, and Director of Production Bioinformatics at the Icahn School of Medicine at Mount Sinai, New York. An award-winning expert who has worked on many successful drug development projects, his current research focuses on generating treatments for diseases, with patents secured for his work in both oncology and immunology.

Bioinformatics and Computational Biology in Drug Discovery and Development

Edited by

William T. Loging, PhD

Associate Professor of Genetics and Genomic Sciences
Director of Production Bioinformatics
Icahn School of Medicine at Mount Sinai
New York, NY

CAMBRIDGE
UNIVERSITY PRESS

University Printing House, Cambridge CB2 8BS, United Kingdom

One Liberty Plaza, 20th Floor, New York, NY 10006, USA

477 Williamstown Road, Port Melbourne, VIC 3207, Australia

314-321, 3rd Floor, Plot 3, Splendor Forum, Jasola District Centre, New Delhi - 110025, India

79 Anson Road, #06-04/06, Singapore 079906

Cambridge University Press is part of the University of Cambridge.

It furthers the University's mission by disseminating knowledge in the pursuit of education, learning and research at the highest international levels of excellence.

www.cambridge.org
Information on this title: www.cambridge.org/9781108461153

First published 2016
First paperback edition 2018

A catalogue record for this publication is available from the British Library

Library of Congress Cataloging in Publication data
Name: Loging, William T., editor.
Title: Bioinformatics and computational biology in drug discovery and development / edited by William T. Loging.
Description: Cambridge, United Kingdom; New York: Cambridge University Press, 2016. | Includes bibliographical references and index.
Identifiers: LCCN 2015041412 | ISBN 9780521768009 (hardback)
Subjects: | MESH: Drug Discovery–methods. | Biomarkers, Pharmacological. | Computational Biology–methods. | Data Mining–methods.
Classification: LCC RS420 | NLM QV 745 | DDC 615.1/9–dc23
LC record available at http://lccn.loc.gov/2015041412

ISBN 978-0-521-76800-9 Hardback
ISBN 978-1-108-46115-3 Paperback

Contents

Contributors

Mark Crawford
LS Pharma – Retired

Thomas B. Freeman
Capella Biosciences, Inc.

Jeff Handler
JAH Associates LLC

Marilyn Lewis
Pfizer Inc.

William T. Loging
Icahn School of Medicine at Mount Sinai

Pek Lum
Capella Biosciences, Inc.

Roy Mansfield
Pfizer Inc.

Jonathan Phillips
Boehringer-Ingelheim Pharma Inc.

Raul Rodriguez-Esteban
Roche Inc.

Telmo Silva
ClicData RCS

Yirong Wang
Mount Sinai Genetic Testing Lab
Connecticut Icahn School of Medicine at Mount Sinai

Bryn Williams-Jones
Connected Discovery Ltd.

Ke Xu
Bristol-Myers-Squibb Inc.

Foreword: The future of drug discovery and healthcare

The era of big data is transforming healthcare

The world we currently live in is unprecedented with respect to the rate of technology expansion and application of these technologies across a hierarchy of health-related problems. Businesses are generating more information, acquiring more data, and aggregating more data on customers than ever before, with the hope that the appropriate analytics applied to these data can inform on all aspects of their product's performance and aid in evolving products and in better targeting of products to specific customer groups. Just as climatologists use big, comprehensive data to predict weather patterns, and quantitative traders on Wall Street use Big Data to assess when to buy and when to sell stocks, so the medical field will soon harness that same power of big data to better understand patient populations, when to treat, and how to treat. However, today, very large-scale data and the integrative analytics that are applied to such data hold no greater promise perhaps than for the healthcare industry. The application of high-throughput informatics technologies to biological problems exemplifies the disciplines of Computational Biology, Bioinformatics, and more recently Clinical Genomics. When one views recent achievements following the hard-won successes of the Human Genome Project and the amount of genomic data being generated on populations of individuals to better understand phenotypic variation in these populations (from disease to the evolutionary architecture of human populations), it is clear that computational tools, advanced algorithms, and a diversity of highly skilled personnel are required to make the most of these data in order to deliver on the promise of the post-genomic era: improved diagnosis, treatment, and prevention of disease.

Understanding disease in a drug discovery setting

Human health is the product of many interacting, deeply complex systems: a person's physiology, their genome and other molecular features, the state of many different types of cells in their body, of their organs, their physical environment, the bacteria and viruses in that environment, their activities, their social interactions, and their medical care. Those who will prevail in the future of developing the most efficacious and safe interventions to treat and prevent disease will be those who can obtain the best data, perform the most informed analyses to identify the patterns associated with disease and wellness, and build the most predictive models based on those patterns – where others may see only noise.

The typical drug discovery process spans 10–15 years, requires investment of roughly one billion dollars, as well as a great diversity of skill sets mainly engineered around the idea of designing a single molecule or biologic to activate or suppress the activity of a single protein. Drug discovery researchers over the last many decades have adhered to an overly simplified view of disease that is very much in need of a revolution in mindset, experimentation, and in approaches and workflows. A large gap often exists between informatics scientists and biologists, which very often limits the potential for advancement of our understanding of complex disease processes. To achieve the ultimate success in developing the most efficacious and safe interventions to treat and prevent disease, an intimate partnership must exist between the informatics and biology experts. Understanding informatics approaches to Big Data is only part of the answer; the application of these approaches and how they interface to solve the problems of delivering drugs to market are critical components as well. Several chapters in this book illuminate this. Often, management and biologists are not aware of the need for advanced computational approaches that can consider millions of variables of information simultaneously to better inform decision making, what exactly the computational approaches can deliver, what is required to enable this delivery and the limitations of such approaches. In the future, the companies and organizations that successfully integrate computational biology approaches with basic biomedical research and drug development will be best positioned to succeed in the quest of improving the lives of patients and well individuals. A major goal of this book is to help bridge these divides in perspective and understanding.

Laboratory bench to patient bedside

The healthcare industry is rapidly changing – payers, payees, providers, patients, and consumers more generally are taking control of data for use in healthcare outcomes, disease assessment and understanding, prediction, as well as health and lifestyle choices. Successful companies will seek to understand how to use these data to make more informed, more accurate, more precise decisions regarding the health of the individual, delivering the right intervention for any disease condition to the right patient at the right time (Schadt *et al.*, 2005), what we now refer to as precision medicine. Only recently have we had the technology to collect very high-dimensional (hundreds of thousands to millions of features), diverse complex molecular data on each individual in a cost-effective manner, and to analyze these data to understand individual variation in disease processes and responses to drugs and even wellness, where identifying features that have enabled individuals to buffer disease, live longer, healthier lives, promises to have the highest impact on overall human wellness (Friend and Schadt, 2014).

Healthcare data are coming from a growing spectrum of sources: electronic medical records (including family medical history), imaging data (MRI, PET scans), genetic information (whole genome sequencing of DNA or whole transcriptome sequencing of RNA), metabolite, protein, lipid and sugar information generated from tissue samples, pathology reports (on tumors), blood tests, urine tests, and so on. However, one of the more dramatic Big Data generators to explode onto the scene over the past several years are the network-enabled wearable and implantable devices that empower individuals to monitor extensively their own health, with devices such as the Apple Watch, FitBit, Nike FuelBand, and HexoSkin enabling a wide array of measurements that can be collected longitudinally, from activity levels to heart rate, blood pressure, and pulse oximetry. Increasingly, these types of devices and mobile apps run on smartphones are generating continuous, longitudinal data on individuals that can provide a far more comprehensive assessment of an individual's health compared to the extremely limited snapshot that gets taken in the doctor's office perhaps once per year. In the not so distant future, more information relating to the health of individuals will exist outside of their doctor's office than inside. Those biopharmaceutical organizations who can make use of these data, in both a clinical as well as a preclinical setting, will be well-placed for success in the future. That level of success will be marked by the number of life-saving therapies that are brought to market or the number of interventions that increase overall well-being in the human population.

Given the wide range of data types and the sheer scale of these data, the challenge is how best to collect them, to organize them in systematic ways, to effectively manage them, compute on them and mine them, so that we can derive meaningful insights across vast numbers of patients that ultimately will inform on tailored treatment paths.

Adaptive drug trials will evolve the standard of care

To get to this future of precision medicine, we will need to transform the way in which interventions are clinically validated and approved by regulatory agencies. We will need to run the type of adaptive trials that can identify and validate a new standard of care as evidence accumulates. For example, as more and more data are generated on patients, these data will be stored in big databases across any number of medical centers, with access to these data shared across medical centers. This information can then be connected to powerful supercomputers that crunch the data on many hundreds of thousands of patients to help predict the exact type of disease a given patient has, the molecular features that best define what that type of disease involves, and from this, identify the most appropriate treatments and/or clinical trials, including drugs, that in turn are communicated to the treating physician. As the physician makes decisions based on this knowledge to treat the patient (the physician is not obligated to follow the predictions made from the big data, but rather the predictions made help engage the physician's mind with the most complete data to recognize patterns relating to the patient condition), the response of the patient can also be assessed, whether their condition improved, got worse, required additional treatments, and so on. Based on this feedback, which is also entered into the database, the models can be updated in an iterative fashion based on the accuracy of their predictions, and in the end the empirical studies will show that the model better predicts treatment path and as a result becomess adopted as the standard of care. In this way, the patient population becomes the clinical trial population, and the clinical trial employs advanced dynamic, adaptive, deep-learning procedures to maximally consider all data over time to reach the most accurate and validated conclusions.

To enable this new world of healthcare, we see now rising a new generation of biologists who are being trained simultaneously in traditional biomedical sciences, as well as in informatics and statistics. It is these new biologists that will help drive revolutionary progress in the biomedical sciences and in healthcare. I sincerely hope that the approaches and technologies described in this book will inspire this next generation of scientists on how to engage this emerging consumer-centric model of healthcare, to think and dream big about the future of healthcare and drug discovery, and in so doing positively impact the human race by enhancing our well-being.

Eric E. Schadt
Jean C. and James W. Crystal Professor of Genomics
Chairman and Professor, Department of Genetics and Genomic Sciences
Founding Director, Icahn Institute for Genomics and Multiscale Biology
Icahn School of Medicine at Mount Sinai in New York

References

Friend, S. and Schadt, E. Translational genomics. Clues from the resilient. *Science.* 2014;344:970–972.

Schadt, E., Sachs, A., and Friend, S. Embracing complexity, inching closer to reality. *Sci STKE.* 2005;295:pe40.

Acknowledgments

A person is often the sum of experiences and the people one meets. That said, I have had the privilege of working with many amazing scientists and leaders throughout my time on this planet. To thank each of these people in depth would take an entire book in itself. However, I will try to provide brief credits. To the subject matter contributors of this book, for their contributions to the field and their work on constant revisions of both text and figures in their endless search of perfection; to members of Cambridge University Press, for their professionalism and patience on timelines. For my current and past team members; it has been said that if you want to be happy, surround yourself with people you enjoy working with. To my PhD mentor, Dr. David Reisman, for teaching me how to dig deep in the art of molecular sciences. For the departed Prof. Geoffrey Zubay (whom I never met) for inspiring me on how a textbook should be written and the inner workings of life. To Drs. Anton Fliri and Robert Volkmann for their mentorship and revealing to me how drug discovery works. To Dr. Lee Harland and fellow contributor Bryn Williams-Jones, for their friendship and long discussions of data science and the Edwardian era. Finally, and most importantly, to my family: Christie, Bri, and Liam, for their support, as well as their sacrifice, in supporting my path in the life sciences.

1 The art and science of the drug discovery pipeline: History of drug discovery

William T. Loging

The disease condition is a standard in the lives of humans worldwide. As far back as the Roman era, early investigators pondered the reason for a wide array of disorders, such as typhoid and polio. In no place was this more evident than the bubonic plagues of the Middle Ages, which are reported as causing the deaths of more than 30% of the population in Europe (Alchon, 2003). As the observational science of biology grew, individual scientists increased their understanding of the underlying causes of human illness. In the nineteenth century, scientists like Louis Pasteur made significant contributions to the human understanding of microbiology and bacteriology. Pasteur's determination to understand human illness was born of the fact that several of his children did not survive to adulthood (Feinstein, 2008), a standard occurrence preceding the advent of twentieth-century medicine. Less than 200 years ago, prior to Pasteur's discoveries, it was coarsely thought that life spontaneously generated from inert materials (Farley and Geison, 1974); this thinking gave little value to the washing of hands and other hygienic procedures. However, additional discoveries quickly followed, such as those made by innovative physician scientists like Joseph Lister, who deduced ground-breaking procedures on aseptic treatment of patients.

The fundamentals of modern drug discovery can be found as an outline to the pioneering work of both Edward Jenner and Alexander Fleming. Although occurring more than 50 years ago, their contributions to combating human illnesses have collectively saved tens of millions of human lives; work that first started with an observation – one that caught their interest (Willis, 1997). Jenner observed that milkmaids were less susceptible to smallpox, and hypothesized that their immunity was due to contracting cowpox, an illness less virulent but related to smallpox. He successfully tested this hypothesis by inoculating human test subjects with cowpox and observed that this protected them from contracting smallpox (Barquet and Domingo, 1997). In the 1700s, it was estimated that nearly half a million people in Europe were killed by the Variola virus, the cause of smallpox (Behbehani, 1983). Due, in part, to Jenner's pioneering work on smallpox vaccination and immunity,

Bioinformatics and Computational Biology in Drug Discovery and Development, ed. W.T. Loging. Published by Cambridge University Press.

the World Health Organization declared in 1979 that smallpox was eradicated worldwide.

The path to Alexander Fleming's discovery of antibiotics began when he observed *Penicillium notatum* fungi growing in laboratory glassware (that he was about to wash after being on vacation), and noted the mechanisms of antibiotic zones. Fleming was not thinking about a "target" in the way that modern drug discovery scientists think, but he merely noticed that bacteria were not growing where some should have been. It is interesting to note that Fleming could have let this observation pass without notice, but luckily for the millions of people whose lives were saved by his discovery, he exhibited one of the main characteristics of an innovative scientist: he was inquisitive about what he observed. In fact, during his observation he was famously quoted as remarking "That's funny" (Brown, 2004). The road to commercial antibiotics, ones that patients could orally take to cure their bacterial infections, was not an easy one, but it did serve as a template for modern-day drug discovery. The creation of the drug discovery paradigm that was created in the twentieth century relied on inputs from multiple scientists as well as high-level strategic thinkers (Paul *et al.*, 2010). However, in the example of both Jenner's and Fleming's discoveries, once the impacts of their findings were noted as beneficial in alleviating the effects of human disease, it was reasonable to assume that treatments for other illnesses could be identified by applying similar innovative approaches. The idea that one could create a treatment that would cure or alleviate diseases instilled hope for other debilitating illnesses like polio and led to subsequent additional discoveries. Because of this, generations will never have to fear the words "smallpox" or "polio"; it is a fact that we too often take for granted in today's modern world.

The birth of computational biology

Biologists are often at a disadvantage when it comes to understanding the complex workings of the human body. Unlike engineering or other scientific fields, no complete blueprint exists for the complex, everyday workings of the human condition. One of the discrete advantages to our race is that we are not clones of each other; however, genetic variations often make it extremely difficult to account for a singular working model of a human being. This fact forces biologists to "learn as they go along" and requires them to rely on comparative biology for understanding the inner workings of normal and diseased states. Comparative biology is conducted by first observing a normal population of cells, tissues, or even a live organism and comparing their functions to those with disease. During the molecular biology revolution of the 1980s, scientists began generating more and more data about cellular machinery by using technics such as Southern blotting (Southern, 1975) and phage display. As these data were being generated, the standard biologist was becoming quickly inundated with higher amounts of information to sift through. The advent of gene sequencing also meant that researchers now had to measure, store, and review long lists of CTAGs, the makings of the genetics code. A new field of science needed to be created in order to keep up with these large volumes of data. Fortunately for these researchers, the 1980s was the beginning of the computer age and created a serendipitous intersection in the two fields for which both biology and computer science came together in the form of bioinformatics and computational biology. Whether described as dry biology, *in silico* biology, conceptual biology or knowledge discovery (Weeber *et al.*, 2003), or electronic biology (eBiology; Loging *et al.*, 2007), bioinformatics and computational biology techniques have a key role in the application of electronic data in drug discovery.

The fields of bioinformatics and computational biology are, by nature, deemed to be high-throughput for a number of reasons. First, they make use of multiple well-known, large-scale and open-source data that are coupled with a wealth of established uses and literature. This lends naturally to the second reason: they are not often dependent on initial "wet-lab" investigations before they can begin. Third, protocols and workflows created by users within these categories are comprehensive enough in that they can often be repurposed to address demands from multiple teams investigating different diseases, rather than having to be applied to one distinctive field of study. This makes it possible to reuse the time and investment already created, along with the user knowledge in setting up these computational workflows across numerous drug discovery projects. Last, many of the pipelines described in this book can be rerun regularly, which provides a means for apprising new data and therefore continuous new findings. It is necessary to point out, and is often a reason why the promise of computational approaches can often be left unfulfilled, that these methodologies still eventually require input from human disease scientists. However, the features described within this book provide a broad overview to introduce the novice, as well as the experienced computationalist, to methods that can be implemented to provide a higher rate of return on investment, when compared to other, less high-throughput styles of computational biology used within drug discovery. Such practices can become a useful element to the *in silico* workbench of any drug discovery organization, as it is often noted that there are far more research and development projects to support than there are computational scientists. The subsequent chapters of this book will follow these forms of computational biology application along the drug discovery pipeline; from target identification, to small-molecule identification and optimization and ending with the evaluation of the physiological effects of a candidate drug in the clinical phases.

The drug discovery pipeline as assembly line

Biochemist Akira Endo is credited with the discovery of the statin, a class of drugs that has been shown to prevent cardiovascular disease. Investigating how one might combat high-cholesterol phenotypes in patients with familial hypercholesterolemia, following in Fleming's line of thinking, he hypothesized that fungi generate small molecules to protect themselves against other opportunistic organisms. He further suggested that inhibiting cholesterol synthesis of the invading species could provide a selective advantage to the fungi, because it was known that fungi cell membranes contain ergosterol in place of cholesterol. Endo's research proved correct as he discovered the first statins from his studies on fungi (Endo *et al.*, 1976). The statin class went on to generate billions of dollars in revenue for the companies that market it. Endo never benefited financially from his discovery, despite the statins being among the most commonly prescribed medications worldwide (Simons, 2003). The sheer impact of Endo's work was highlighted by Nobel Prize Biochemists Michael S. Brown and Joseph Goldstein: "The millions of people whose lives will be extended through statin therapy owe it all to Akira Endo" (Landers, 2006). In 2005, sales of the statin class were estimated at $18.7 billion in the USA, of which atorvastatin was listed as one of the world's best-selling pharmaceuticals in history (Simons, 2003). I mention Dr. Endo's work for two reasons: first, to again document the tremendous impact that a single innovative scientist can have on the lives of millions of people, and second, to show that the success of Dr. Endo's approach subsequently led others to envision that such discoveries could be created *en masse*. As the statin class began generating billions in revenue, additional emphasis was placed on new discoveries, but the path to generate these life-saving drugs can be a long and arduous process; therefore, the mainstream drug discovery pipeline was relied upon to generate novel discoveries.

The term "drug discovery" itself is somewhat of a misnomer, as the vast amount of drugs that are brought to market are more often made than they are discovered. The historical paradigm of drug discovery processes, popularized from the early 1990s until today (or the writing of this book), has its foundation in the following:

Stage 1 Target Identification: Disease area scientists employ comparative biology methodologies to define a therapeutic target (which is often a protein that plays a role in the illness). The approach can range from obtaining a target idea from a publication to conducting complex protein or pathway screens within *in vitro* environments.

Stage 2 Lead Identification: The identified target is formulated into a functional high-throughput assay and screened against large chemical/biological libraries, often composed of millions of potential candidates, for identifying probable small molecule or antibody inhibitors.

Stage 3 Lead Optimization: Any "hits" from the lead identification stage are then passed through an optimization phase for assessment of the chemical and/or biologic states that make the candidate drug-like. Lead candidates are then adapted, modified, and improved upon in order to progress the drug frontrunner to the preclinical safety and efficacy testing.

Stage 4 Preclinical Assessment: If the candidate meets all the criteria to pass previous stages, it is progressed into rounds of preclinical safety and expanded efficacy assessments. This stage is often, depending on the disease and possible organism model, where the candidate is administered to lower forms of mammals, such as rodents or non-human primates.

Stage 5 Clinical Phases: Lastly, the potential drug is used in human clinical applications, starting with safety testing Phase I with a small group of human subjects. If successful, the candidate is then moved into continued efficacy testing in actual patients who are affected by the disease, in a smaller (Phase II) and subsequently broader population (Phase III). Success is measured in how safe the drug is, as well its level of efficacy, in human patients.

Figure 1.1 An example of the standard drug discovery pipeline.

Idea to targets

New drug targets are often ascertained from vast amounts of work conducted mainly in the field of biology. As mentioned before, these data provide investigators with a specific area of intervention that would affect the progression of a disease. This is highlighted where several research organizations have used different methods, whereas some researchers obtained target ideas by solely reading scientific publications. As stated before, when the concept of target identification became more prevalent, scientists began relying on non-publication-based methods that utilized the large amounts of data generated from methods such as SAGE (serial analysis of gene expression) and other leading techniques (Velculescu *et al.*, 1995), such as genes chips, and then to next-generation sequencing (NGS), where we currently stand today. New target work and comparative biology often go hand in hand, and no other field has benefited more from this marriage than the field of oncology (Jiao *et al.*, 2010). Standard experimental design collates normal, as well as cancerous, tissue that is measured from two separate states using several of the techniques which will be reviewed in this book. The results often lead researchers to potential proteins that may play a role in the disease; for example, oncogenes that control cell growth are often found to be overamplified in the cancerous experimental arm compared to normal, whereas tumor suppressors are often under-represented. Once researchers have a protein of ascertainable function and it has been verified through public knowledge or internally validated work, then researchers can search for molecular inhibitors (small molecules or proteins) that can inhibit the activity of this protein, therefore bringing the system back into balance with the hopes of curing disease. In the early 2000s, sequencing of genetic material was considered "high-throughput" as 30,000 base pairs could be processed in a time frame of about 3 months. In fact, when the human genome project was completed in 2001, it took 10 years to complete, as well as costing more than a billion dollars (Collins and Galas, 1993). Consider now the current speed at which genetic material is sequenced using technology like that of NGS. These applications can be applied to comparative biology studies of both genes (DNAseq) and gene transcripts (RNAseq). For example, Illumina now markets their HighSeq X Series NGS machine as having the ability to sequence the entire genome of a human being in less than one week, with cost ranges in the area of \$1000 (www.illumina.com). This rapid advancement in genome sequencing technology, where speed is increased and cost is reduced, is nearly unprecedented within human history progression and has even been presented to surpass that

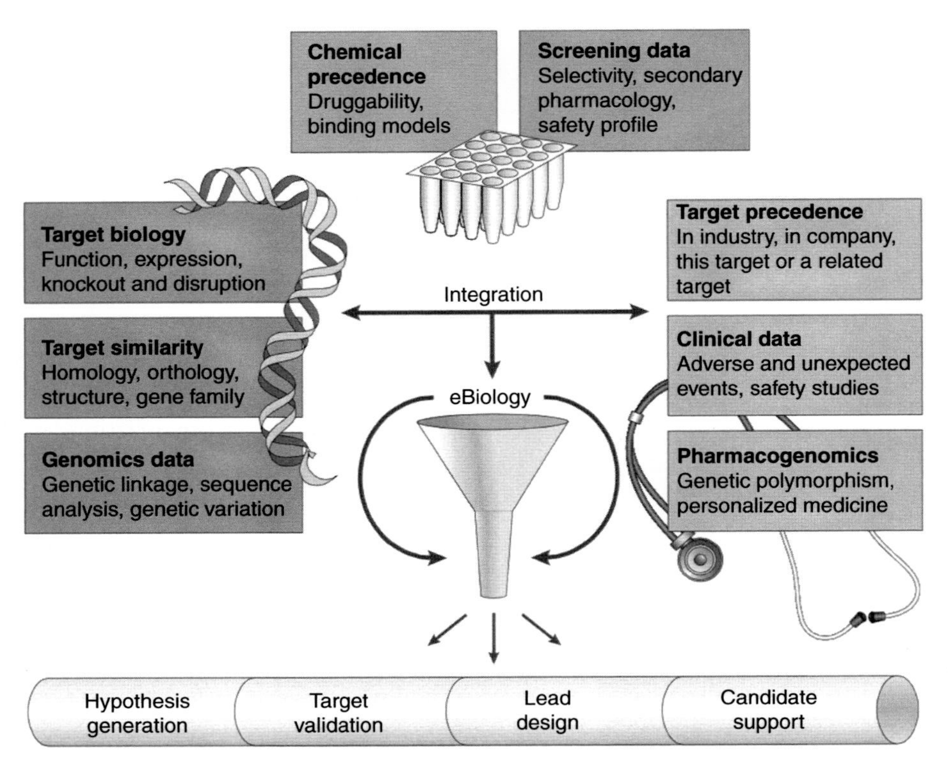

Figure 1.2 Breakdown on DDP by phase. From Loging et al. *Nature Reviews Drug Discovery* 2007.

of Moore's Law. With the large volumes of data that are generated from such high-capacity experiments, computational biology approaches will continue to be in demand for the drug discovery scientist.

Lead identification

Once a single, or multiple set, nominated target protein has been identified, the target is moved to lead identification, where suitable small-molecule (or antibody) inhibitors are identified through a biological drug design assay and the use of large chemical libraries. The target is often placed in a high-throughput set of experiments, often referred to as "screening," where millions of small molecules are interrogated for their ability to inhibit the function of the target protein in a reporter assay. Lead identification is a key decision-making stage for the research organization conducting it and can easily generate billions of data points, which require significant computational resources performed by both computational chemistry as well as computational biology experts. The druggability of the protein target via a small-molecule or antibody inhibitor is essential for the success of the discovery program. Computational biology plays a major role in the large-scale analyses of these new protein sequences and their druggability by structural homology searching (Lui and Rost, 2002), as well as data interpretation support from computationally driven docking experiments of small molecule candidates. Molecular modeling techniques are typically applied to this area, as these approaches include nuclear magnetic resonance (Hajduk *et al.*, 2005) and x-ray crystallography (An *et al.*, 2004). Computational biology pipelines can provide immediate impact such as by reviewing more than 18,000 crystal structures from the Protein Data Bank. Again, those researchers who can link together computational pipeline aspects will be those who are most successful; for example, Cheng and colleagues (2007) created a technique for a druggability calculation that joins together multiple approaches that generate a number of molecular descriptors into a single algorithm. These methods use established data from orally available drugs, along with correlation-specific data such as ligand molecular mass along with protein-binding pocket surface area; key physicochemical properties are then linked together with structural analyses in order to obtain a scoring of druggability. Not only can this approach be applied specifically to this example, but it can also be run computationally across the large sets of crystal structure data, providing a high-throughput approach to target assessment. As mentioned earlier, a small-molecule approach may not be the only desired endpoint of a drug discovery program; antibody drug generation discovery projects also benefit from such methodologies through employing additional data such as epitope placement, cellular placement and plasma availability of proteins. Comparable to chemical approaches that often use large libraries of small molecules, the antibody engineering provides additional methods of drugging a specific protein that have led to the identification of drugs that were ultimately approved for use in humans.

Lead optimization

As a candidate drug progresses through the DDP, researchers must focus on the off-target selectivity effects, in addition to focusing on the binding of the molecule to its primary target. Naturally, a number of *in silico* approaches have been created that utilize additional data that go beyond protein homology searches. For example, investigators have utilized not only chemical data, but also genomic information to address kinase selectivity (Birault *et al.*, 2006). Further, *in silico* techniques can utilize added data such as predicted *in vivo* drug metabolites, and assist in understanding the physiological effects of known drugs. Whereas the lead optimization phase is characterized as making a drug candidate more drug-like, how this is accomplished is dependent upon several established approaches, such as the analysis of the structure–activity relationship (SAR) data. SARs are the features by which changes in the drug candidate may affect its ability to bind to and modulate the activity of its target protein. Again, intuitive scientists continue to rework the established drug discovery paradigm, in which researchers proved utility in the analysis of large SAR databases (Nettles *et al.*, 2006). One of the scientific areas that

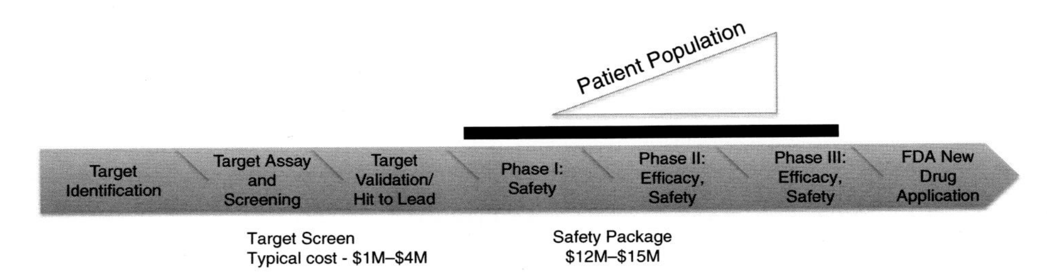

Figure 1.3 The clinical phases.

drug discovery researchers earnestly pursue is the understanding of small-molecule off-target activity.

A straightforward process would be to screen the entire human proteome with the compound in question; however, economic and technological requirements would make such an approach unfeasible. However, an *in silico* protocol that utilizes a broad panel of the druggable genome that is representative of major protein families can provide the interaction potential of the small molecule in question. This illustration panel of proteins is not limited to just the primary target, and creates a profile "fingerprint" of pharmacologic activity for a given compound. Fliri *et al.* (2005) provide a practical demonstration of how a druggable proteome cross-section can be analyzed to produce a pharmacologic fingerprint for a small molecule. These patterns can then be compared to a larger screening drug test set that numbers in the thousands. In fact, the database need not be confined to drugs; a wide array of additional molecules, such as natural products, can also be used to obtain comparative pharmacologic data. These data, termed biospectra profiles, have also been shown to be useful in functional activity prediction (agonist versus antagonist). This is useful because the small molecules brought out of high-throughput screening in lead identification do not normally provide this information (Fliri *et al.*, 2005). Biospectra can also lead researchers to understand how physiological effects are associated within secondary pharmacology of existing drugs, an approach that naturally lends itself to drug candidate disease repositioning (Campillos *et al.*, 2008). These approaches will be discussed in a relevant case study later in this book.

Preclinical and clinical phases

Perhaps the most critical stages of the drug discovery process are the progression through the preclinical and clinical phases before finally being submitted to appropriate governmental agencies for approval. As mentioned earlier, the candidate will likely enter into *in vivo* testing for the first time in this phase, being dosed to a wide array of mammals and even higher organisms, such as non-human primates. This is often where a critical hand-off also can take place, as the candidate is passed from the research side of the organization to the development side, which is responsible for the actual creation of the candidate in large-enough qualities as required within human-based trials in a clinical setting. This critical area has recently been under the spotlight for its strategic placement within the drug discovery process and has led to the generation of a new conceptual field referred to as "translation medicine." Many researchers had already been formulating opinions on the methods required to "translate" the drug candidate from the research arm to the development and clinical stages, and the information needed to make such a transition successful. Do the safety and efficacy noted in non-human models fully extrapolate into what will happen when the drug reaches the clinic? What data are required early on in the process in order to predict success? These are several of the areas of focus for the translation medicine scientist, whose main role is to provide scientific assurances that a specific drug program will ultimately be successful in the clinic – and therefore obtain approval by the appropriate governmental agencies. Several computational biology tools and methodologies have been created to specifically assist the translational medicine scientist in dealing with the large amounts of disease-based information. These data span a wide array of genomics, as well as even EMR (electronic medical records) that allow for meaningful comparisons between those of normal and diseased patient populations. Often, a drug candidate is found not to be efficacious in the clinic – then what? Many companies park their frontrunner in a Phase II or Phase III graveyard where the candidates often languish, never to be utilized in experiments, or in the extreme, never even talked about again, within their respective company. In the later chapters of this book, we will review strategies to conduct "repositioning experiments," providing insight into other possible indications in which researchers can test their candidates. Prior to the year 2000, researchers appeared to focus more on the primary indication that brought the program forward with little thought of conducting repositioning experiments. However, repositioning projects success stories, such as Pfizer's sildenafil, proved that effectively replaced drug candidates could generate billions of dollars in revenue. In the early 2010s, additional computational approaches – as well as entire companies devoted to the subject – were created (Ashburn and Tho, 2004). Along with examples of successfully placed drug programs, it proves that often the information that may draw connections between drug target and indication can go unnoticed by seemingly expert staff and that computational approaches can provide insight at a level not previously noted within the scientific field (Loging *et al.*, 2011).

An example of when drug discovery works: *PCSK9* and the rise of genomics-era drugs

In the early 2000s, scientists began to report on patients who had a family history of low "bad" or low-density lipid (LDL) cholesterol (Levy, 2015). These serum chemistry levels of LDL did not change regardless of diet or exercise. Comparative

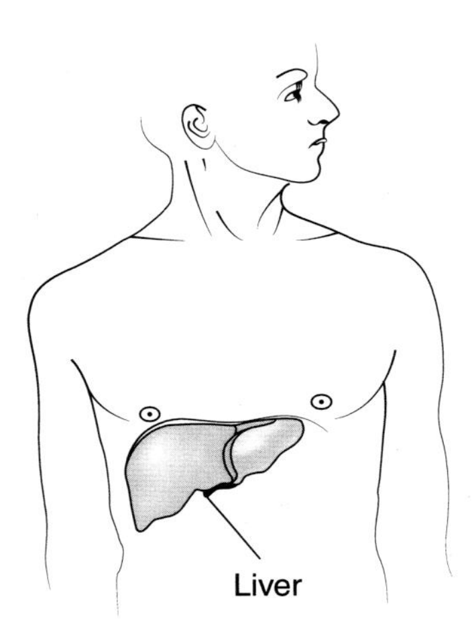

Figure 1.4 PCSK9 protein regulates cholesterol levels in humans.

biology experiments were with these individuals against other groups of patients who had familial high level of LDL. It's interesting to note that these patients with extremely low LDL also had family histories of no existence of heart disease, whereas the opposite was true for those patients with high LDL. This provided an excellent opportunity for comparing the two populations and utilizing the newly introduced approaches of whole-genome screening. It took more than 5 years, but researchers discovered that one of the major differences between the two extreme populations was a gene called *PCSK9* (Jialal and Patel, 2015). Once the gene was identified, the function of the newly associated cardiovascular gene had to be elucidated. By conducting additional computational and wet genomics studies, it was later hypothesized that *PCSK9* regulates the levels of LDL receptors in patients, a process that oversees how the liver cleanses bad cholesterol out of the blood. In these studies, the patients with low LDL contained mutations in *PCSK9* that caused a loss of function in the gene, whereas the patients with high bad cholesterol had carried *PCSK9* mutations, which amplified the function of the coding protein. It is now known that *PCSK9* acts as a protein marker that influences the recycling of the LDL receptor on the liver's surface. Due to the size of the binding interface between the LDL receptor and *PCSK9*, the vast majority of small molecules tested did not provide inhibitory utility, therefore driving researchers to employ biotherapeutic drug discovery approaches that have led to several antibody candidates, which are on track for Food and Drug Administration (FDA) approval by mid to late 2015 (Sabatine *et al.*, 2015). Preliminary data from Phase III studies of anti-PCSK9 antibodies has given clinical signals of not only reducing LDL levels but also reducing the chance of heart attack or stroke in those patients administered this drug candidate (Bloom *et al.*, 2014). The reason why I bring this point up is the very important observation that drug discovery should not be limited to the direct applications of the step-by-step pipeline, such as idea to target first, as highlighted in the example of *PCSK9*. The pipelined process was reversed slightly

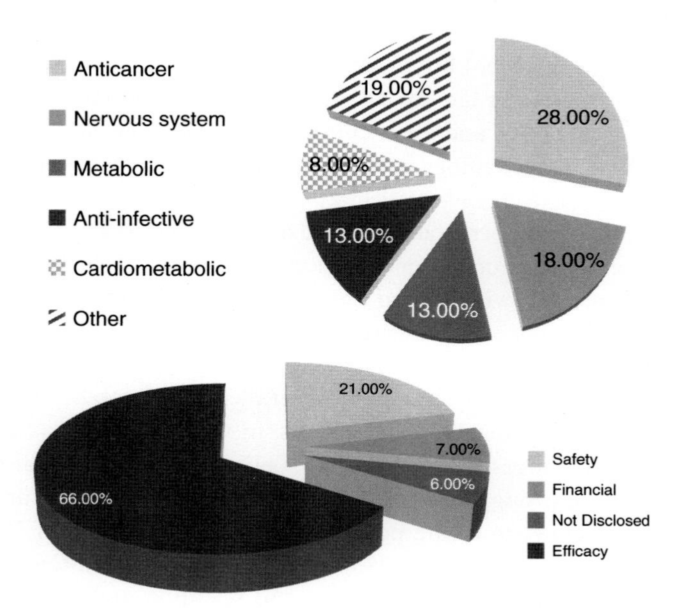

Figure 1.5 Reasons for drug discovery failures.

as the positive phenotype was first observed in human patients, and then computational/laboratory methods and utilizing genomics data from identified individuals allowed for the isolation of the mechanisms that led to a positive outcome in those patients. An actual example of this way of thinking has been promoted in a new undertaking called the Resilience Project that hopes to expand upon such drug discovery successes like *PCSK9* (Friend and Schadt, 2014).

Does drug discovery have an Achilles' heel?

As Moore's Law has expanded the computational utility of scientists in the fields of drug discovery and biotechnology, it has become commonplace for multiple publications, as well as news reports, to highlight a novel procedure that will potentially open the floodgates for finding new drugs to cure disease. However, the question remains: what has been our collective success rate? A retrospective analysis of recent candidate submission failures in the drug discovery field (Arrowsmith, 2011) has brought forward several interesting observations. First, a large number of programs that lacked prior knowledge perquisites, often an important consideration for success, have subsequently taken place within the expanded realms of the "genomics era." This means that the majority of these programs did not have a large amount of biological understanding prior to the start of the project. In a perfect world, drug discovery would be somewhat like a cookbook process. One puts in a measurable amount of effort, a specific volume of time, and if everything goes well, there should be a positive amount of success obtained. However, unlike a recipe in a normal kitchen, this "cookbook" contains very subtle nuances that relate to the biological information for the project, as well as unknowns concerning the new chemical or biological entity, and all of this in an age where perhaps not all of the mechanisms of these biotechnology cooking materials are known (Arrowsmith and Miller, 2013).

Second, a multitude of informatics tools and databases, containing large amounts of internally generated genomics information, have been applied very early on within projects, sometimes with limited impact and success when it pertains to the actual drug project, leaving many researchers to ask: what approaches are we missing? The real problem lies in the direct application of the computational workflows to the problems that are being questioned. A parallel of this can be noted with the molecular biology field in the 1970s and 1980s, in which substantial numbers of *in vitro* tools and techniques were emerging. Agarose gels existed well before the advent of Southern blotting; however, it took the emergence of composite protocols for such workflows to create powerful procedures like that of PCR (polymerase chain reaction; Mullis *et al.*, 1986). As these were directed towards specific biological questions, it helped to rapidly expand scientific understanding of the systems studied; an effect that we are still filling today within the "genomics age." In a similar way, composite workflows in computational biology might help to generate tangible value from the many *in silico* resources that are being placed for use directly into a drug discovery setting.

Here we also highlight the publicly available data sets (see Appendix II) which emphasize the fact that an immense amount of data exist and will continue to be generated from approaches such as high-throughput sequencing, gene expression studies and even patient medical records. It will be up to those individual scientists who can conceptualize the large amount of data that exist and how they can be utilized within the connected workflow to find value and bring scientific understanding. Multiple examples exist where successful completions of these pipelines have made significant contributions to each of these stages, and where these projects delivered drugs that were administered to humans to alleviate their respective disease. The following chapters of this book will review in detail aspects of several of those examples.

While the subject of drug discovery could easily be implemented into a year-long academic course at any major academic institution, this book is meant to introduce the reader to two major points that often go unconnected. Those being that the science of drug discovery and the computational tools that are often employed need to work in unison for successful progression of the drug program. From myself and the additional authors who have contributed here, we provide an overview of the current state of affairs within the pharmaceutical space – it is hoped that as research scientists and students review the information contained within, they may come up with additional or novel approaches to speed up the process of drug discovery, and in doing so provide alleviation of human disease and its effect on humankind, which is paramount and the main focus of individuals who aspire to go into the field.

Focus on production computational biology

One of the more necessary requirements to conducting these informatics activities is rooted not so much in the scientific approach, but in the overall strategy of how it is conducted. Granted, the science is obviously very important to the success of any work in drug discovery; however, it is additionally important to highlight that informatics be conducted with a sense of business processes or accomplishment using the four rules for success. What are business processes? These are related to first defining objectives or setting goals. (Figure 1.6) (1) What needs to get done as well as what do you want to achieve? (2) Also, are you able to set realistic timelines of when you expect the work to be completed? Without concrete timelines, an individual's work can often drag on within any real delivery. (3) Define the resources needed for such work. (4) Are you able to identify and prepare for any and all roadblocks that might inhibit delivery? Roadblocks could hide any aspects of your work.

Defining goals is a very critical point, as it must be done with specifics in mind. In order to adequately describe the objective, the timeline that is involved must be described as well as that in which they will expect to achieve it. One can imagine that if you're trying to run bioinformatics within a business environment this work needs to be defined for delivery in a timely manner. Too often, research pertaining to exploratory concepts may have the luxury of not having specific timelines associated. Computational biology that is being conducted on the industrialized scale should be protected from the distractions that can exist when areas of interest present themselves (Murray, 2011). Even publishing a paper in a prestigious journal can be a distraction. This does not preclude the fact that researchers need to be focused on their delivery, but also speaks to the concept of time allocation and how one

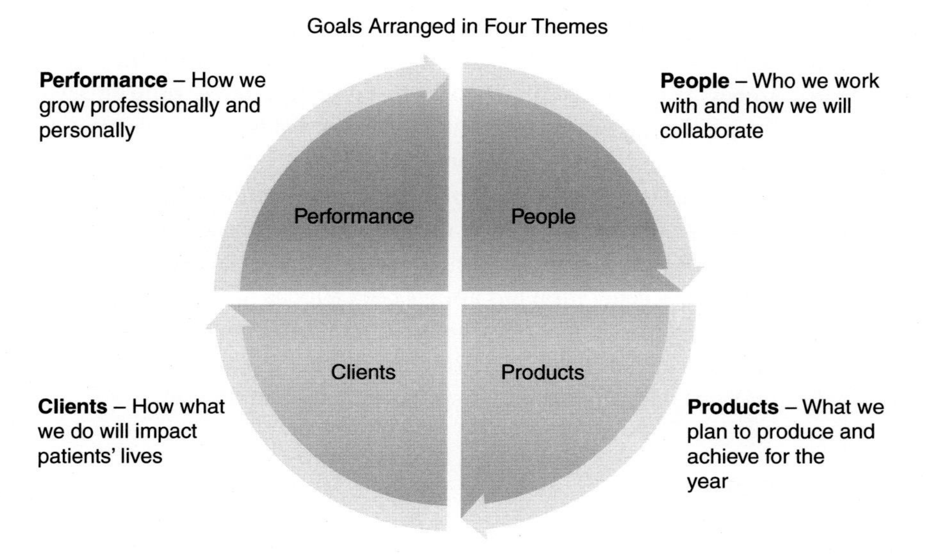

Figure 1.6 The goals creation process arranged into four sub-themes.

arranges their day with project work as well as how one perceives their success in that work, i.e., the Dunning–Kruger effect (Kruger and Dunning, 1999). Participation in discussions with one's mentor/management can also help in the alleviation of trying to do too many things at once. This is why within the art and science of the drug discovery paradigm, there must exist aspects to deliver results from data, in a timely manner, that run like clockwork – as these strategic approaches also have their place. In the end, it often comes down to the leadership of the drug discovery organization and their ability to provide a strategic landscape to their colleagues (Murray, 2011). Those leaders that can organize and distribute the resources required for the efficient running of bioinformatics and computational biology will be aligned for success within the other moving parts of the drug discovery process.

As previously outlined, the innovation process for discovering drugs is an intricate and laborious process that often can succumb to multiple pitfalls. Many a researcher in the field has labored for years, and at the end of their career never fully worked on a project that was ultimately approved for use in humans. However, I have been fortunate in my career in having contributed to programs that are currently approved for use in humans and extending and saving lives. In fact, the drug label of Jardiance® (empagliflozin; recently approved for use in humans; https://www.jardiance.com/) actually contains a section to which my team and I contributed. Many a computational professional has asked me about the process by which we were able to contribute to these programs and my feedback to them can be summed in two words, "vision" and "passion." If you can constantly think from a strategic standpoint about how the computational approaches can be dovetailed with the biology and the needs of the program, then you will be well-positioned for success. The passion not only for the science, but also for the people who you work with, as well as the lives of those you are trying to impact, also blend to innovation and imitation, two other qualities that are required for success, not just within drug discovery but any business. I and my fellow authors hope that by illustrating some of the means by which *in silico* approaches can be utilized, they can inspire other researchers to create new approaches that directly address problems within drug discovery. Lastly, it is essential to appreciate that fields such as computational biology must function as fixtures in the presence of sound (biological) experimental design. Without such focus on the experiment and the questions such approaches are trying to answer, the process at best will only provide erroneous findings. It is important to point out that merging novel technologies within established approaches to scientific hypothesis generation has always been a focus of the pharmaceutical field. The authors of this book envision that computational biology will continue to remain an essential element to the field of drug discovery and empower the discoveries of future drugs that will alleviate the effects of human disease.

References

Alchon, S. A. *A Pest in the Land: New World Epidemics in a Global Perspective*. Albuquerque, NM: University of New Mexico Press, 2003.

An, J., Totrov, M. and Abagyan R. Comprehensive identification of 'druggable' protein ligand binding sites. *Genome Informatics*. 2004;15:31–41.

Arrowsmith, J. Trial watch: Phase III and submission failures: 2007–2010. *Nature Reviews Drug Discovery*. 2011;10:87.

Arrowsmith, J. A decade of change. *Nature Reviews Drug Discovery*. 2012;11:17–18.

Arrowsmith, J. and Miller, P. Trial watch: Phase II and Phase III attrition rates 2011–2012. *Nature Reviews Drug Discovery*. 2013;12:569.

Ashburn, T. and Tho, K. Drug repositioning: Identifying and developing new uses for existing drugs. *Nature Reviews Drug Discovery*. 2004;3:673–683.

Barquet, N. and Domingo, P. Smallpox: The triumph over the most terrible of the ministers of death. *Annals of Internal Medicine*.1997;127(8 Pt 1):635–642.

Behbehani, A. B. The smallpox story: Life and death of an old disease. *Microbiology Review*. 1983;47(4);455–509.

Birault, V., Harris, C. J., Le, J., *et al.* Bringing kinases into focus: Efficient drug design through the use of chemogenomic toolkits. *Current Medicinal Chemistry*. 2006;13:1735–1748.

Bloom, D. J., Hala, T., Bolognese, M., *et al.* A 52-week placebo-controlled trial of Evolocumab in hyperlipidemia. *New England Journal of Medicine*. 2014;370:1809–1819.

Brown, K. *Penicillin Man: Alexander Fleming and the Antibiotic Revolution*. Stroud: Sutton Publishing, 2004.

Campillos, M., Kuhn, M., Gavin, A. C., Jensen, L. J. and Bork, P. Drug target identification using side-effect similarity. *Science*. 2008;321(5886):263–266.

Cheng, A., Coleman, R. G., Smyth, K. T., *et al.* Structure-based maximal affinity model predicts small-molecule druggability analysis. *Nature Biotechnology*. 2007;25:71–75.

Collins, F. and Galas, D. *A New Five-Year Plan for the United States: Human Genome Program*. Bethesda, MD: National Human Genome Research Institute, October 1, 1993.

Endo, A., Kuroda, M. and Tsujita, Y. ML-236A, ML-236B, and ML-236C, new inhibitors of cholesterogenesis produced by *Penicillium citrinium*. *Journal of Antibiotics (Tokyo)*. 1976;29(12):1346–1348.

Farley, J. and Geison, G. L. Science, politics and spontaneous generation in nineteenth-century France: The Pasteur–Pouchet debate. *Bulletin of the History of Medicine*. 1974;48(2):161–198.

Feinstein, S. *Louis Pasteur: The Father of Microbiology*. New York, NY: Enslow Publishers, 2008.

Fliri, A., Loging, W. T., Thadeio, P. and Volkmann, R. A. Analysis of drug-induced effect patterns linking structure and side effects of medicines. *Nature Chemical Biology*. 2005;1:389–397.

Friend, S. H. and Schadt, E. E. Translational genomics. Clues from the resilient. *Science*. 2014. www.genome.gov/11006943

Hajduk, P. J., Huth, J. R. and Fesik, S. W. Druggability indices for protein targets derived from NMR-based screening data. *Journal of Medicinal Chemistry*. 2005;48:2518–2525.

Jialal, I. and Patel, S. B. PCSK9 inhibitors: The next frontier in low-density lipoprotein lowering. *Metabolic Syndromes and Related Disorders*. 2015;13(3):99–101.

Jiao, Y., Yonescu, R., Offerhaus, G. J., *et al.* Whole-exome sequencing of pancreatic neoplasms with acinar differentiation. *Journal of Pathology*. 2014;232(4):428–435.

Kruger, J. and Dunning, D. Unskilled and unaware of it: How difficulties in recognizing one's own incompetence lead to inflated self-assessments. *Journal of Personality and Social Psychology*. 1999;77(6):121–1134.

Landers, P. How one scientist intrigued by molds found first statin. *Wall Street Journal*. 2006, Jan 9.

Levy, E. Insights from human congenital disorders of intestinal lipid metabolism. *Journal of Lipid Research*. 2015;56(5):945–962.

Loging, W. T., Harland, L. and Williams-Jones, B. High-throughput electronic biology: Mining information for drug discovery. *Nature Reviews Drug Discovery*. 2007;6(3):220–230.

Loging, W. T., Rodriguez-Esteban, R., Hill, J., Freeman, T. and Miglietta, J. Cheminformatic/bioinformatic analysis of large corporate databases: Application to drug repurposing. *Drug Discovery Today: Therapeutic Strategies*. 2011;8(3–4):109–116.

Lui, J. and Rost, B. Target space for structural genomics revisited. *Bioinformatics*. 2002;18:922–933.

Mullis, K., Faloona, F., Scharf, S., *et al*. Specific enzymatic amplification of DNA in vitro: the polymerase chain reaction. *Cold Spring Harbor Symposia on Quantitative Biology*. 1986;51:263–273.

Murray, D. K. *Plan B: How to Hatch a Second Plan That's Always Better Than Your First*. New York, NY: Simon and Schuster Publishers, 2011.

Nettles, J. H., Jenkins, J. L., Bender, A., *et al*. Bridging chemical and biological space: 'Target fishing' using 2D and 3D molecular descriptors. *Journal of Medicinal Chemistry*. 2006;49:6802–6810.

Paul, S. M., Mytelka, D. S., Dunwiddie, C. T., *et al*. How to improve R&D productivity: The pharmaceutical industry's grand challenge. *Nature Reviews Drug Discovery*. 2010;9(3):203–214.

Sabatine, M. S., Guigliano, R. P., Wiviott, S. D., *et al*. Efficacy and safety of evolocumab in reducing lipids and cardiovascular events. *New England Journal of Medicine*. 2015;372(16):1500–1509.

Simons, J. The $10 billion pill. *Fortune Magazine*. 2003, Jan 20.

Southern, E. M. Detection of specific sequences among DNA fragments separated by gel electrophoresis. *Journal of Molecular Biology*. 1975;98:503–517.

Velculescu, V. E., Zhang, L., Vogelstein, B. and Kinzler, K. W. Serial analysis of gene expression. *Science*. 1995:270(5235):484–487.

Weeber, M., Vos, R., Klein, H., *et al*. Generating hypotheses by discovering implicit associations in the literature: A case report of a search for new potential therapeutic uses for thalidomide. *Journal of the American Medical Information Association*. 2003;10:252–259.

Willis, N. J. Edward Jenner and the eradication of smallpox. *Scottish Medical Journal*. 1997;42(4):118–121.

2 Computational approaches to drug target identification

Thomas B. Freeman and Pek Lum

Introduction/overview

In this overview we will examine drug target identification, a key early step in the drug discovery and development process and a significant and interesting challenge. Today, target identification is akin to solving a very difficult puzzle using all of the information that we can generate or have available to identify those points of intervention within a highly interconnected dynamic system that will improve a disease state. I see our current approaches as surrogates for a true understanding of the molecular basis of disease, which when achieved should clearly reveal points for therapeutic intervention. Before we attempt to discuss methods used for target identification, we will begin by exploring what a target is and what hurdles a potential target must overcome to be successfully "drugged." We will examine how the targets from currently approved drugs have been elucidated. A brief historical perspective on target identification will focus on how advances in our understanding of biology have driven our progress. Finally, we will review current directions in selecting drug targets and in understanding disease in an integrated and multi-scale manner.

Drug development is a long and intricate process but one that is quite predictable. The development of a drug usually begins with the selection of a molecular target for modulation by a small molecule or biological compound with the goal of safely improving a disease phenotype. The target concept can be logically extended to include plant biology and perhaps synthetic biology. Crop improvement by increasing yield, quality (i.e., amino acid composition), and insect or drought resistance require identification of points for molecular intervention to bring about the desired phenotype. Similarly, synthetic biology projects require the manipulation of existing biological systems or ultimately creation of modified biological systems, and require a very high level understanding of the genetic and molecular processes involved in producing the desired outcome without destabilizing the system. Our review here will focus on target identification for the purpose of drug discovery and development to address unmet medical need.

Bioinformatics and Computational Biology in Drug Discovery and Development, ed. W.T. Loging. Published by Cambridge University Press.

Drug discovery research is performed at many levels: in the pharmaceutical industry, in biotechnology companies, government laboratories, research foundations, increasingly in computation-based companies, and even larger generic drug companies. In terms of dollars and number of drugs approved, the pharmaceutical industry has been the largest sector in drug discovery to date. Pharmaceutical companies are typically organized around therapeutic areas that, logically, reflect the way diseases have traditionally been defined and the manner in which medical care is delivered (i.e., cardiovascular & metabolic, immunology & inflammation, diabetes, central nervous system). Each therapeutic area maintains a portfolio of projects built around specific targets or molecular mechanisms believed to be important for the pathophysiology of the diseases of interest to that therapeutic area.

Projects within a therapeutic area are organized into drug development pipelines that represent milestones in the advancement of projects through various stages of development based on increasing evidence that the target is important for disease and that modulation of the target will positively impact the disease phenotype of interest. As projects progress through the pipeline they compete with other projects for resources such that the "best" projects advance. The stages in the pipeline begin with exploratory development and advance through stages that include biological assay development, compound screening, preclinical development, and clinical development (including Phase I, II and III development). The general strategy in early drug development is to establish a goal for a certain number of successful proof-of-clinical-concepts (Phase II successes or demonstration of efficacy in a disease cohort) and to build a pipeline of projects sufficient to deliver this number of successes accounting for project attrition along the pipeline. Projects are halted or "killed" during the progression through the discovery pipeline for many reasons, such as safety or toxicity issues, inadequate exposure, lack of demonstration of positive effects in model systems or lack of feasible experimental path forward. Thus, there is a constant need to develop new target ideas to keep the pipeline supplied with active projects. Target selection is a key factor in the success of a drug development program. When successful it typically requires about 15–20 years from initiation of an exploratory discovery project to market approval, which means that drugs in late-stage clinical development today reflect project decisions made in the late 1990s. Since that time, the economic climate and tolerance for risk in the pharmaceutical industry has changed dramatically. This is reflected in many mergers and acquisitions, dramatic reductions in numbers of research scientists, research budgets and a narrowing of therapeutic area portfolios. Unfortunately, much of this narrowing of scope has impacted disease indications that are of major importance as the global population expands and ages.

What is a target?

Before discussing methods for drug target identification, it is important to explore what is required for a molecular drug target to be successfully "drugged." We will begin by looking at the requirements for drug approval and conditions that help differentiate between a successful and a failed drug target. In a later section we will look into the actual targets that have been successfully drugged and the historical track record of how the drug against these targets were developed.

What are the requirements of a target for which drugs have been successfully developed? Safety and efficacy are the minimum requirements for a successful drug. These are established through two major regulatory milestones in drug development in the United States: the IND (Investigational New Drug application) and the NDA (New Drug Application), both granted by the FDA (Food and Drug Administration). The IND is obtained prior to the start of clinical development, and the application requires detailed information in three key areas: preclinical (animal) pharmacology and toxicology; chemistry and manufacturing information; and clinical protocols and investigator information. The purpose of the IND application is to demonstrate that human research subjects will not be subjected to unreasonable risk from compounds that have never before been administered to human subjects. The NDA is an application requesting that the FDA approve a new drug for sales and marketing. A major portion of the NDA is demonstration of substantial evidence of drug efficacy demonstrated by means of controlled clinical trials. We will look at safety and efficacy and other factors influencing the success of drug development in a bit more detail.

Safety is a prime consideration for successful drug development. Careful assessment of toxicity is conducted during preclinical development. This is a critical and rapidly evolving field in drug discovery. Potential toxicity can be caused by the targeted molecular mechanism, or can be caused by "off-target" effects of a candidate compound, meaning the candidate compound binds to and influences molecules in addition to the desired target causing unwanted side effects. Mechanism-based toxicity likely indicates that modulation of the chosen target is unsafe and will likely cause termination of that particular program. Target selection at exploratory development must carefully consider any evidence of potential safety issues and these must be monitored throughout preclinical development. Safety and tolerability in healthy human subjects (as well as drug candidate pharmacokinetics) are demonstrated in carefully controlled Phase I clinical trials typically involving healthy human volunteers. Safety is monitored in patients with disease in later-stage clinical trials and beyond. Reporting of adverse events is mandatory at all stages of development and post-approval.

Drug efficacy requires demonstration of evidence of clinical benefit beyond the current standards of care including symptomatic relief, delaying or preventing progression of disease or, ideally, curing disease. Many of our current therapies are palliative or provide symptomatic relief without curing the disease; this is the case for many drugs for treatment of depression, for example. Symptomatic relief alone has improved the quality of life for many millions of patients and has brought much financial success to drug development companies and their shareholders. Delaying disease progression is often a therapeutic goal, as in the case of Alzheimer's disease where a significant delay in onset or progression would improve the quality of life, or in chronic kidney disease where a delay in progression would lengthen the time before start of dialysis or need for a kidney transplant. These seemingly modest goals would have a significant impact on individual lives and on healthcare delivery, but remain beyond our grasp for the moment. It is an interesting observation that some of our most successful classes of drugs do not treat disease *per se*, but rather treat risk factors for disease, as in lipid-lowering and anti-hypertensive drugs. This is possible based on the results of ongoing decades-long publicly financed studies such as the Framingham Heart Study begun in the late 1940s. Studies such as these have analyzed the risk factors, health and outcomes of thousands of patients for decades, allowing for clear demonstration that reduction in risk factors such as low-density lipid (LDL) cholesterol levels prevents cardiovascular-related deaths. It is a daunting task to try to identify a single molecular target to modulate with a drug molecule in a person with disease to slow, reverse or cure this disease. We will explore these methods later.

We must look more deeply at the factors that influence successful drug discovery as it bears on drug target selection, because several factors have changed over the past several decades in ways that raise the bar that must be met for successful drug approval. These include both the successes and failures of drug development over the past several decades and changes in global demographics and healthcare economics.

The phenomenal success of the pharmaceutical industry and of drug development over the past 60 years has addressed the needs of most people with many major medical conditions with fairly efficacious and fairly safe drugs that were developed and typically approved against placebo, the standard of care at the time. Diseases such as hypertension, hypercholesterolemia and hyperlipidemia, major depression, diabetes, various forms of cancer and rheumatoid arthritis are a few examples of diseases for which multiple blockbuster drugs have been developed. This success has led to an expectation among pharmaceutical executives and investors that regular development of such blockbuster drugs should be the norm and has led to an approach to research and development informed by this expectation. However, today many of these conditions and others are treated with safe and relatively efficacious generic drugs, such as metformin for type 2 diabetes and beta-blockers, such as propranolol, for hypertension. New drug candidates directed at these conditions must show efficacy against or in addition to these current standards of care with a similar or better safety profile. This significantly raises the bar for drug development and presents a significant challenge which may best be met by understanding the molecular mechanisms of the current drugs in the context of the disease such that new drugs can be developed against targets that can act via additive or synergistic molecular mechanisms; drugs developed that meet these criteria will be better drugs and as safe or safer for patients than the current standard of care. We shall see that there is very much opportunity and need for new drugs, including drugs that would be considered blockbusters if developed. This will likely require deeper research into the pathophysiology of targeted diseases, which is increasingly enabled by technological developments in tools, methodologies and support infrastructure.

In addition to acute safety, the long-term safety of drugs, in particular for drugs designed to treat chronic conditions, such as hypertension or diabetes, is of paramount importance in that these conditions require lifelong drug treatment under our current treatment paradigms and has been highlighted by recent Phase III failures and drug market withdrawals. Inhibition of cholesterol-ester transfer protein (CETP) as a target for anti-atherogenic elevation of high-density lipid (HDL) cholesterol level was suggested by genetic analysis of patients with extreme phenotypes – those with very high HDL cholesterol who were found to have genetic variants in the CETP gene that rendered them deficient in CETP activity accompanied by increased HDL cholesterol and apolipoprotein A and reduced LDL cholesterol and apolipoprotein B with no evidence of premature atherosclerosis (Inazu *et al.*, 1990; Morabia *et al.*, 2003). Currently, two CETP inhibitors are in active clinical trials (Evacetrapib (Lily) and Anacetrapib (Merck)). A third CETP inhibitor (Dalcetrapib, Hoffmann-La Roche) was terminated in Phase III trials due to lack of clinical impact (plaque reduction) despite significant elevation of HDL cholesterol levels. Pfizer's Phase III trial and development of the CETP inhibitor, torcetrapib, were terminated due to a significant increase in deaths in the torcetrapib plus atorvastatin arm versus atorvastatin alone. This toxicity has been attributed to "off-target" effects that increased circulating mineralocorticoids and increased blood pressure independent of CETP inhibition (Tall, 2007; Forrest *et al.*, 2008). Such off-target effects can be difficult to predict, but screening studies such as CEREP's BioPrint analysis may help identify off-target binding events for follow-up, especially before the start of clinical trials (Krejsa *et al.*, 2003).

Post-approval controversies related to the cardiovascular safety issues of the thiazolidinedione class of insulin-sensitizing agents has led to withdrawal or use restrictions of members of this otherwise efficacious class of drugs. Voluntary withdrawal of Merck's Vioxx from the market and warnings on the use of Pfizer's Celebrex following concerns about increases in cardiovascular deaths also highlight this issue. Effective risk mitigation in drug development requires the prediction of long-term safety issues in patients or subsets of patients who will be required to be on medications for chronic conditions. Advances in our ability to predict such long-term effects will require increased development of our models of disease at multiple scales: cellular, tissue, organ and organ system level and crosstalk between these levels. Concerns about long-term safety of medicines that require chronic treatment have resulted in more stringent requirements for long-term morbidity and mortality studies for drugs in these therapeutic areas. Such studies are justified given our current limited understanding of the bases of these safety risks; to protect public health requires lengthy and expensive studies that may deter some from working in these disease areas.

It has become increasingly apparent that drug development must consider the heterogeneity of disease and patient individual genetic variation. Many diseases having a similar clinical presentation and clinical diagnosis may be different at the molecular level and therefore may not respond in the same way to a particular therapeutic intervention. Stratification of patients into subgroups based on the molecular disease subtypes will be key to understanding and to effectively targeting each subtype. Similarly, individual patient genetic variability may influence the effectiveness of particular therapeutic interventions. Pharmacogenomics screens are often performed on patients requiring treatment for psychiatric disorders to help guide the choice of therapeutic intervention, dramatically shortening the time until an effective treatment is obtained compared with a more trial-and-adjustment approach. Polymorphisms in two genes (*VKORC1* and *CYP2C9*) play a significant role in response to the anti-coagulant warfarin. Patients prescribed warfarin are regularly tested to help regulate their exposure to the medicine because too high levels cause bleeding and too low levels increase the risk of thrombosis and thromboembolism. Similarly, it is becoming the standard of practice in oncology to perform molecular profiling of tumor (somatic) mutations to determine the best treatment options for the individual patient, as several therapies are only efficacious in specific molecular subsets of patients with a particular cancer.

While outside of the realm of scientific inquiry, commercial viability has become an important factor in determining the success of a drug program. This puts a premium on understanding the molecular context of a target identified for modulation for an indication and understanding how it is differentiated from the standard of care, other therapeutics for the clinical indication of interest and from competing projects within a company's development pipeline. Several years ago it was predicted that the anti-coagulant market would be the biggest drug market by 2016, and in the interim period several new anti-coagulants have entered the market but have not achieved predicted market penetration against the cheap generic warfarin. This has likely had a significant impact on the financial projections of the affected companies.

With respect to target selection, there are two broad categories of drugs to consider: new molecular entities (NMEs) and follower drugs. A new molecular entity is a drug that modulates a target that has never had a prior drug approved and represents a new molecular mechanism and typically addresses an unmet medical need. A follower drug is a drug in development or approved for use that modulates a molecular target in a disease or indication for which another drug is already approved. Follower drugs are very common and typically result from the concurrent development of drugs by several pharmaceutical companies. This reflects accumulation of sufficient data connecting a particular target with a disease leading to the initiation of development programs at multiple companies in a similar time frame. Follower drugs must compete against one another in the marketplace, often resulting in the identification of a best-in-class molecule that limits profitability for the other drugs in this class, resulting in reduced return on investment for many of the competing companies (many markets are large enough to support several successful competitor drugs). However, this often provides alternatives within drug classes, which may be of benefit to subgroups of patients for which the best-in-class drug is not optimal. For example, there are many statins on the market – each with different pleiotrophic effects and different tolerability among patients.

In addition to the many "me too" drug successes there are also many "me too" drug development failures – targets concurrently under development by multiple companies that fail due to lack of safety or lack of efficacy. Unfortunately, in the past these data were often not thoroughly examined and almost never made public or shared. This represents a great loss of potential understanding, a high level of redundant activity and investment and exposes clinical trial patients to similar risk without benefiting from prior experience with similar candidates against the same target. Many drugs have been repurposed for new indications or uses often after successful Phase I studies demonstrate safety followed by Phase II efficacy failure for the intended indication (Loging *et al.*, 2011). It is now often common practice to perform indications discovery exercises for programs in development to identify additional indications for a drug candidate to expand the economic potential of the drug, mitigate risks in event of Phase II failure for an intended indication, or to identify a more streamlined clinical path to first approval.

It is illustrative to look at reported clinical development shortfalls to understand the reasons for and the magnitude of the impact (Table 2.1). From 2008 to 2010 there were 108 reported Phase II failures; of those reporting reasons for failure, 51% reported lack of efficacy, 29% were for strategic reasons and 19% were reported to be due to clinical or preclinical safety issues. Approximately 30% of the failures reported for strategic reasons involved precedented ("me too") targets (Arrowsmith, 2011b; Figure 2.1). At later stages in the development pipeline, combined successes in Phase III and submission have fallen to approximately 50%, with 83 failures between 2007 and 2010. Ninety percent of the failures across all therapeutic areas are due to lack of efficacy (66%) or safety issues (21%). Failures are due to lack of efficacy against placebo (66%), against active control (5%) and as add-on therapy (29%) (Arrowsmith, 2011a). Considering the cost in dollars and other resources, this late-stage attrition is unsustainable and the current drug discovery model needs to be modified. Under the current drug discovery paradigm, we are not able to mitigate significant risk early on in drug development, and we do not learn as much as we should from our development failures (Schadt *et al.*, 2009).

Despite the success of the past, and despite the current high failure rate, significant unmet medical need currently exists and is expanding on several fronts. Most of our successful therapeutics leave subsets of patients with partially or completely unmet needs; efficacy sometimes decreases with time in others; and many of our current treatments are palliative rather than curative. There are many "orphan" conditions for which no drugs have been developed because they have been considered to represent too small an opportunity to be profitable for pharmaceutical companies to pursue (an aspect of the blockbuster model of drug discovery). Patient advocacy and recent government incentives have promoted increased activity directed toward "orphan" diseases. This is reflected in the recent increase in approvals for drugs for orphan indications. New opportunities arise as the global population grows and ages – obesity and diabetes, diseases of ageing such as Alzheimer's disease represent significantly

Table 2.1 Key definitions

Active pharmaceutical ingredient (API) – the biologically active ingredient in a pharmaceutical drug.

Analoging – creating series of structurally similar chemical compounds, with the goal of understanding the relationships between chemical structure and biological activity and pharmacological properties.

Attrition – gradual loss of projects in a drug development pipeline due to issues such as safety or toxicity concerns, lack of efficacy or lack of experimental path forward.

Blockbuster drug – a drug that generates more than $1 billion of revenue per year.

Causality – analysis of statistical associations between variation in DNA, expression traits and phenotypic traits to infer likely directionality of expression and phenotypic trait.

Data-driven – knowledge or hypotheses derived from patterns in data rather than from preconceived ideas or hypotheses.

Etiology – cause of a disease or pathology.

High-dimensional data – data sets that contain very large numbers of measurements.

Hypothesis-driven – approaches derived from prior knowledge and preconceived ideas about what is important in a system.

Multi-scale – data sets that contain data collected across various levels of organization (cell, tissue, organ system, populations) or across time points.

Omics data – data collected on various components of the biological system: RNA, protein, genetic variation, metabolites.

Pleiotrophic effects – gene affects multiple, seemingly unrelated endpoints.

Scraping data – automated retrieval of information from websites.

Secondary metabolites – organic compounds produced by an organism that are not directly involved in the normal growth, development, or reproduction of an organism. Secondary metabolites are often involved in defense against predators.

Synthetic biology – modifying or designing biological systems for a particular purpose.

Text analytics – automated retrieval of high-quality information from text sources.

increasing burdens that need to be addressed. New diseases also emerge as current diseases are subdivided into molecular subtypes. In many indications the opportunities for new therapeutics in terms of potential markets for drugs that show a significant improvement over standard of care represent potential "blockbuster" opportunities. To advance drug development to meet these standards and in particular in indications such as Alzheimer's disease and major depression, where our current understanding and utility of preclinical models are limited, will require significant advances in our understanding of these diseases. In order to meet these existing and growing medical needs in a productive and sustainable manner, the current model of drug discovery needs to be examined and reimagined. Munos (2009) performed a retrospective analysis of drug approvals covering the period from 1950 to 2008 and found that the rate of introduction of new drugs of novel mechanism into the market has been essentially constant throughout this 60-year period, despite dramatic increases in pharmaceutical research and development investment (Figure 2.2). One interpretation of this observation is that the amount of research and data collection is increasing dramatically but analyses and productive use of these vast data are lagging.

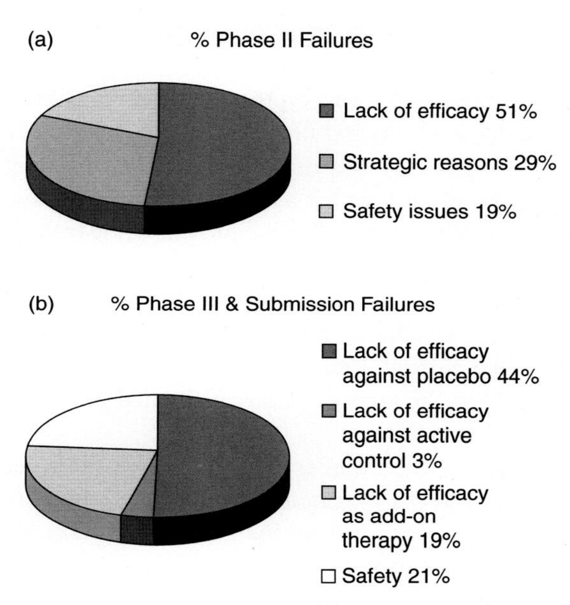

Figure 2.1 Late-stage clinical failures. A significant number of late stage clinical trials have failed over recent years. (a) There were 108 failures in Phase II between 2008 and 2010. (b) There were 83 failures in Phase III clinical trials and submissions between 2007 and 2010. Greater than 50% of the failures in each class are attributed to lack of efficacy in the targeted disease population. Figure is based on statistics reported by Arrowsmith (2011a, 2011b).

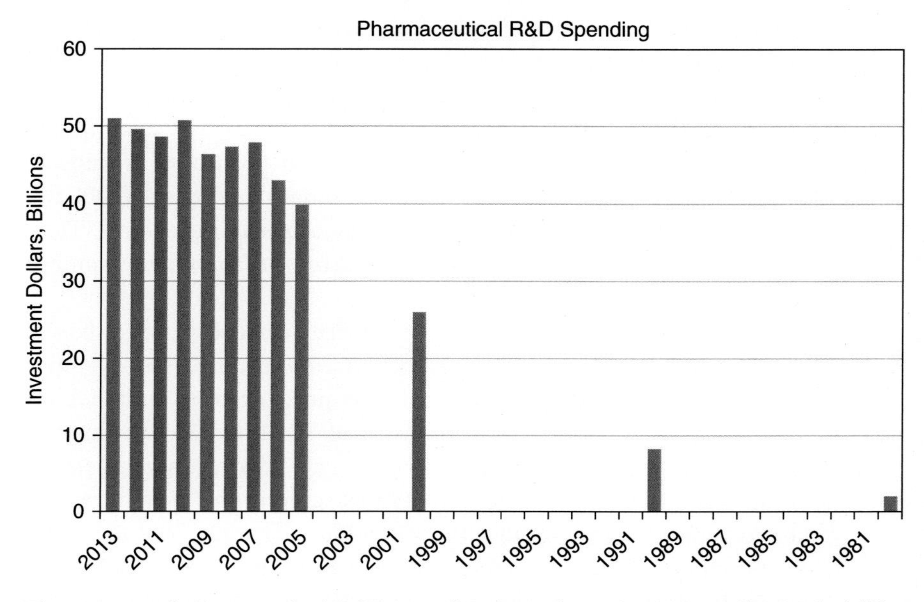

Figure 2.2 The amount of pharmcuetical R&D spending from the early 1980s until 2013 (in billions of dollars) (from Munos, 2009).

How have we identified the currently drugged targets?

Studying drug development failures informs on the effectiveness of these processes relative to the inputs. Analysis of successfully drugged targets may suggest how these have been identified. In 2014, 41 new drugs were approved, an increase over recent years and the second highest year for approvals of new chemical and biological entities (Saboo, 2015). About half of the approvals came from big pharma, the other half coming from smaller pharma, biotech and generic companies. This is interesting given the proportionately larger investment by big pharma. Many of the new approvals were for smaller or formerly "orphan" diseases, addressing important needs; many of these drugs approved for orphan indications are unprecedented targets. About half of the approvals have estimated peak sales at $1 billion or greater.

There is no hard estimate of the number of targets that have been successfully drugged. Defining the target of a drug is a difficult task. In many instances the actual target may not be known; a drug may have a primary target identified but may interact with other proteins and molecules. The number of proteins and other molecules with which a drug interacts throughout the cell may be very large. Resources such as Drug Bank enumerate drug–target relationships based on literature reports (Wishart, 2006). Literature estimates of the number of successfully drugged targets range from about 300 up to approximately 1500 (see Drews and Ryser, 1997; Drews, 2000; Overington *et al.*, 2006). Typically these analyses use the "gene" as the unit of target. Assuming 25,000 genes in the human genome, successful drug targets comprise 1.2%–6% of the genome. The majority of targets that have been drugged include receptors (45%), enzymes (28%), hormones and growth factors (11%); other important classes are ion channels and nuclear hormone receptors (Figure 2.3). There are about 30,000 entries in the FDA Orange Book of drugs; however, many of these are directed at the same molecular target (when the target is known), many vary only in formulation, and several are combinations. These 30,000 reduce down to approximately 7500 unique active pharmaceutical ingredients. There is probably no good way to estimate the number of targets that have been entered into development programs and have failed.

There are four principal strategies for drug discovery: target-based, phenotypic, natural products, and biologics. Target-based screening, especially high-throughput screening, is hypothesis-driven and is the most common approach today, and typically involves screening large libraries of small-molecule compounds against a recombinant target protein in an *in vitro* assay. There are several assumptions inherent in this approach: that the activity measured in this preclinical assay will translate to clinically meaningful activity in a Phase II study; that the molecular form of the target molecule used in screening is relevant to human disease; and that modulating this target will modify the human disease phenotype in the desired manner. These assumptions are typically not thoroughly validated until Phase II or Phase III clinical trials, which might be reflected in the high percentage of failures at this

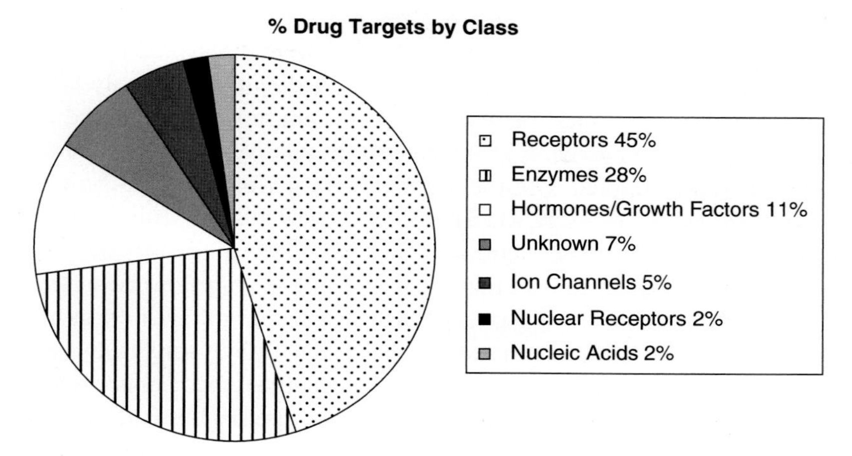

Figure 2.3 Percentage of drug targets by molecular class. The majority of approved drugs as of 2003 were target receptors, especially G-protein-coupled receptors. These are followed by enzymes and hormones and growth factors. Ion channels and nuclear hormone receptors have also been targeted. The figure is based on data reported by Bleicher *et al.* (2003).

late stage. The phenotypic-based approach involves screening molecules against a biological system without preconceived ideas or hypotheses about target or mechanism. The key assumption is that the biological assay system employed (i.e., cell culture, co-culture, tissue, *in vivo*, or organ-on-a-chip) reflects the molecular state that underlies the human disease state. Current research tools must be employed to uncover the underlying target and mechanism of action for candidates that drive the desired phenotype in the assay. Drug discovery by use of or modification of natural products has a long tradition in drug discovery. Many drugs/targets have been and continue to be discovered by this route (e.g., statins). In the natural environment there is an incredibly large diversity of chemical molecules that have evolved to work with or against biological systems; these molecules and the microbes that produce them remain largely uncataloged and are difficult to exploit. In the future, perhaps synthetic biology will help harness some of the incredible chemical and synthetic capabilities present in the secondary metabolite arena. Finally, biologics-based drugs are typically developed by target-based approaches with an alternate screening paradigm.

There have been several published analyses of the historical productivity of drug discovery based on drug approvals by the FDA and information about the approaches used to develop specific drugs. Swinney and Anthony (2011) analyzed 259 drugs approved between 1999 and 2008, of which 183 were small-molecule drugs and 56 were biologic drugs (20 were imaging agents which were not included in the analysis). Focusing on the 75 first-in-class drug molecules, 50 were small molecules (67%) and 25 were biologics (33%); 28 of these first-in-class small molecule drugs were discovered by phenotypic screens, and 18 were discovered by targeted screening. Of the 164 follower or "me too" drugs approved, 83 (51%) were discovered by target-based screening, 30 (18%) were discovered by phenotypic-based approaches, and 31 (19%) were biologics. A disproportionately large number of drugs with novel targets or mechanisms of action were discovered by phenotypic screening. This is especially striking given the relatively small number of phenotypic screens run compared with the number of target-based (high-throughput) screens run each year. Conversely, when considering new follower drugs against targets with validated mechanisms of actions, target-based screening outperformed phenotypic screening; this likely reflects that the underlying assumptions have already been validated in the clinic. Using a target-based strategy for follower drugs on precedented targets is much less risky than for drugs against novel targets, but drugging precedented targets is much less likely to address unmet medical need. Taken together, these observations suggest that target-based approaches can be very productive with validated targets; however, for identification of novel targets, a hypothesis-free phenotypic approach may be more productive than a hypothesis-driven approach. This reflects our limited understanding of the function of a potential target within the molecular context of the disease of interest. The increasing amount and types of data collected and the growing ability to analyze and integrate these data are beginning to allow better understanding of the molecular basis of disease and are leading to better methods for drug target identification.

Drug target identification: an historical perspective

Advances in biological understanding, approaches and tools since the beginning of the twentieth century have driven drug discovery forward, opening new doors of inquiry and new perspectives and at times have held us in place until new tools are developed and new understandings emerge.

In pharma in the last 60 or 70 years of the twentieth century, certainly since World War II, target identification was one in the same with basic research. Initially our understanding of physiology and biochemistry was relatively simpler than that of today. Painstaking low-throughput research in physiology, biochemistry and pharmacology in pharma, in academic and government research labs produced the framework understanding of much of the physiology and pathophysiology on which we build today. This work opened up vast new areas of biology to fruitful investigation, and led to the development of many of the first and subsequent generations of basic medicines that have improved health globally and set the standards of care against which new medicines today must compete. Importantly, this research proceeded at a scale and level of complexity that was very well suited for the intuitive ability of the human brain to understand, analyze, and integrate these data.

Early exploration of the sedative effects of early anti-histamine drugs illustrates how basic or "bottom-up" research has opened up vast new areas of biology and in this case contributed to the development of many of the highly successful first- and second-generation psychiatric drugs that so changed the world of mental health care. In the early 1930s, Laboratoires Rhône-Poulenc began analoging anti-histamine molecules in search of molecules with better properties and synthesized promethazine, a novel anti-histamine molecule with more pronounced anti-histamine and sedative properties than previous molecules. Based on clinical observations of patients treated with promethazine, two French surgeons used the new drug as an anesthetic prior to surgeries (López-Muñoz *et al.*, 2005). The effects were so dramatic that a Rhône-Poulenc chemist synthesized a series of promethazine analogs and screened for compounds with the least peripheral effects. The candidate selected for use was RP-4560 or chlorpromazine, more commonly known as Thorazine or Largactil. Chlorpromazine was released to physicians for testing and was found to be successful as an anesthetic booster and shown to have startlingly dramatic anti-psychotic effects when administered to institutionalized patients. It was released onto the market in 1953 and was widely adopted for the treatment of many psychiatric disorders. Use of chlorpromazine dramatically reduced the populations of institutionalized mental patients around the world and greatly reduced the use of therapies such as electro-shock and insulin-shock for treatment of these patients (Preskorn, 2010a). Several of the clinicians involved in the clinical development of chlorpromazine for psychiatric uses received the 1957 Lasker Award for Clinical Medical Research.

The exploration of the sedative effects of early anti-histamines that led to the development of chlorpromazine in part launched systematic scientific study of chemical therapeutic modulation of psychiatric disorders. Differentiation of the effects of the family trees of drug analogs led to the identification of different drug groups that elucidated neurotransmitter receptors and receptor subclasses. Exploitation and continued development of this knowledge and associated techniques and methodologies has led to the development of generations of highly effective and blockbuster CNS drugs that have transformed the way mental health is managed (Palmer and Stephenson, 2005; Preskorn, 2010b).

In the late 1960s and early 1970s, the elucidation and antagonism of the histamine H2 receptor for treatment of gastric and duodenal ulcers by Sir James Black and colleagues at Smith, Kline & French (now GlaxoSmithKline) is heralded as the beginning of rational drug design. Sir James Black, a Scottish pharmacologist, while at ICI Pharmaceuticals led the team that developed propranolol, the first beta-adrenergic antagonist for treatment of heart disease. It had been observed that histamine analogs and various anti-histamine compounds had divergent effects in various tissues, in particular some compounds led to release of stomach acid while others did not. These observations led to the hypothesis that there must be more than one type of histamine receptor, and Black and his team systematically developed the compounds and bioassay systems to elucidate the pharmacological effects of these compounds. After several years of intensive work they had characterized the H1 and H2 histamine receptors and developed a therapeutically viable specific histamine H2 receptor antagonist, cimetidine, for the inhibition of gastric acid release. Cimetidine is better known by its commercial name, Tagamet. Basic research including observations from clinical, biochemical, physiological, and pharmacological research led to a well-defined and validated molecular target that was readily exploited by the technical prowess of a large pharmaceutical company to produce one of the most successful drug franchises until recent times. Development of cimetadine changed gastric ulcers from a painful, life-limiting and life-threatening condition to one that is relatively easily managed for most people. Sir James Black and colleagues were awarded the Nobel Prize for Physiology or Medicine in 1988 for their achievements in drug development.

During the second half of the twentieth century, the development and expansion of molecular biology have allowed the dissection of biological systems and processed to the finest detail. Thirty-five years ago there were molecular biologists and there were biologists – two fairly distinct disciplines. Today, virtually every biologist has internalized these concepts and tools into routine practice – even to the point where they are readily available to high school students, do-it-yourselfers and other citizen scientists. Rapid expansion of molecular biology catalyzed by the excitement of new ways of understanding biology and competition for research grant funding produced a seismic shift in the center of gravity in biology laboratory research toward a molecular and reductionist approach with a concomitant reduction in more holistic and physiological research activity and training. Molecular biology has changed the nature of biological research in ways that have greatly increased the scale and details of our framework understanding of physiological systems elucidated by earlier generations of biologists. Drug discovery has firmly integrated the tools, philosophy and perspectives of molecular biology into the fabric of research such that expression of proteins, knocking in and out of genes, constructing customized cell lines and model organisms are routine.

Technological development of genome-wide technologies such as transcriptomic profiling (hybridization-based arrays, and later next-generation sequencing (NGS) technology), genotyping, availability of genomic sequences and functional annotations for humans and model organisms, global measurement of other molecular entities (proteomics, metabolomics, fluorescence-activated cell sorting (FACS)) have continued to increase the scale and richness of the data collected that can be used to add to the description of the state of physiological systems. New classes of molecules such as microRNA and other species of non-protein-coding RNA have been elucidated; these newly identified classes of molecules are part of the richness and complexity of biological regulation. Pharma has readily adopted these technologies and is vigorously working on methods to apply these data to drug discovery. NGS technologies (DNA-Seq) have allowed exploration of individual and population level genetic variation. Other applications of NGS technology, such as Chip-Seq, are enabling more fine-level exploration of regulation of gene expression, as exemplified by the ENCODE project. The increasing power and decreasing costs of computational resources, continued development of new and maturation of existing statistical analysis, such as causal analysis and machine learning algorithms, facilitate analysis and integration of the multi-faceted and multi-scale data we are beginning to collect. This is the cutting edge we are now living at in biology, in general, and in drug discovery in particular.

Target identification in pharma today

In conducting laboratory research and reading literature, it is natural for a curious scientist to seek ideas as to how to manipulate the phenotypes or responses under study. Scientists in pharmaceutical company therapeutic area programs are incentivized to actively seek and develop proposals for candidate drug targets to be considered for entry into the exploratory portfolio of projects. The aim is to select from among many proposals for candidate targets those that have the highest probability of leading to successful Phase II proof of clinical concept. For a chronic kidney disease indication this might mean identifying a target that when selectively modulated in patients with later-stage kidney disease has a high probability of safely halting or slowing progression to end-stage kidney disease. While this is the aim, operationally at the stage of entry into the exploratory portfolio, the selection is based on a decision as to which candidate targets have the best package of data at that time to support allocations of technical and financial resources. Data are derived from publicly accessible databases, scientific literature, meeting abstracts and patents, tools such as Ingenuity Pathway Analysis gene report (for each gene an ontology with hyperlinks to citations in a curated knowledge base) and internal gene expression and data repositories. Data presented include evidence of linkage to human disease, indications of safety or toxicity concerns, tissue distribution of expression, sequence homology to related human proteins and those of model organisms and a likely experimental path forward, to name a few examples. This relatively manual target selection process takes advantage of the judgment of the combined diverse experience of a group of scientists, is relatively hypothesis-driven and is highly dependent on existing knowledge.

The underlying concept for this basis for target selection is that each target has a certain quantity and quality of data associated with it at a given point in time. Certain targets (successfully drugged targets, for example) are very well characterized and have likely been tested and exploited as drug targets. Other targets, despite having intriguing pieces of data associated, will have too little characterization to justify allocation of resources against better-characterized candidates. There exists a sweet spot in between these extremes where targets have enough characterization to justify exploratory investment but not so much that the opportunity has passed. Scientists are often assisted in the collection of information by colleagues with expertise in scraping and databasing structured data and in systematically mining literature sources. These processes could be automated and executed on a periodic basis for all potential targets to generate a "target data density trajectory," plotting increases in quantity and quality of data for each target over time. This would require the development and refinement of text analytics queries, algorithms for analysis, quality assessment, and scoring. Such a system would augment the efforts of individual scientists and could highlight targets that are approaching a threshold of data that justifies deeper inquiry.

There is an inherent bias in such hypothesis-driven approaches in that the more information we have about a target the more likely it is to have resources in pharmaceutical development allocated or in grant funding awarded. Edwards and colleagues (2011) performed a historical analysis of kinase research spanning the interval before and after the sequencing and annotation of the human genome. In the period up to 2000 about 12% of now-known kinases were studied, published on, and were subjects of patents. With the elucidation of the whole kinase family following the human genome project, kinase work remained unchanged with the 12% of well-known kinases still comprising the bulk of the kinase grants, literature and patents, despite the fact that many of the orphan kinases were shown to have interesting disease associations based on systematic screening of kinome function by hypothesis-free techniques such as RNAi. The factor that separated the studied from the unstudied kinases was the presence of tool compounds that allowed the generation of enough data to justify investment and grant application funding. The development of novel kinase tool compounds for particular kinases facilitated study of the formerly neglected kinases and has opened up new opportunities in biology. This example highlights the importance of the need to complement hypothesis-driven approaches with unbiased approaches that are not limited by our prior knowledge and will likely open up new opportunities. Collection of data from biological systems on a global scale is facilitating analyses beyond current understanding and hypotheses.

Pharmaceutical therapeutic area scientists work to advance compounds through the discovery pipeline; they also spend considerable effort in developing tools and model systems to further their understanding of diseases of interest and to aid in drug development. Data generated in the development, characterization, and use of these systems continually adds to the baseline of data that is used in selection and progression of candidate targets. Types of biological systems are developed that are thought to reflect or underlie aspects of the human disease phenotypes; these include cell culture (primary, cell lines and co-culture systems), tissue preparations (primary, organ-on-a-chip, for example), animal models of disease (mechanistic and disease recapitulating), and human disease tissue samples when available. The underlying assumption is that to some extent study of these systems can be translated to humans and to the human diseases of interest. In the case of a chronic kidney disease group, cell lines derived from key kidney cell types and sometimes primary cells can be cultured and perturbed; kidney tissue preparations, isolated glomeruli, cortex or medulla, and systems like the glomerulus co-culture system can be studied to give a more integrated view. Animal models like the unilateral ureteral obstruction model, in which the ureter draining one kidney is ligated to induce rapid hydronephrosis and renal fibrosis in the ligated kidney, is a mechanistic model of renal fibrosis. Data can be compared between the ligated and the unligated contralateral kidney, as well as between the ligated and the sham-ligated kidneys. *In vivo* studies using inbred and F1 strains of diabetic rats are used to model human disease.

Phenotypic, biochemical, and transcriptomic changes are among the endpoints for which data are collected depending on the system. Changes are monitored with disease progression in animal models with or without treatment with placebo, standard of care and tool and experimental compounds. Specific phenotypes of interest are monitored in cell and tissue systems. Typically, changes in gene expression and other biochemical measurements (protein expression, protein phosphorylation, cytokine or matrix production) are correlated with phenotypic endpoints (markers of cellular activation, percent glomerulosclerosis, percent tubulointerstitial fibrosis, etc.).

These approaches are hybrid approaches combining prior knowledge of diseases and disease models with endpoints that are thought to reflect the phenotype of the disease with global measures, such as transcriptomics. The challenge is that typically hundreds or thousands of transcripts change with treatments and correlate with changes in other markers and there is often no way to divide these changes into those causal for and those reactive to the phenotype. Therefore, it is difficult to prioritize these potential targets. The commonly available tools for pathway analysis, gene set enrichment and gene ontology (GO) enrichment, facilitate organizing these data into functional categories based on prior knowledge which we recognize as relevant to the disease condition. However, these methods organize only a subset of the total number of changes and therefore probably do not reflect the full richness and complexity of interactions in the experimental system. TALE (Tool for Approximate LargeE) graph-matching is an interesting algorithm for comparing networks across treatment groups or across species (Tian and Patel, 2008). This algorithm has been used to aid in guiding translation at the differential expression transcriptional network level between animal models of disease and human disease sample. For example Berthier and colleagues (2012) constructed transcription networks using human lupus nephritis and living donor kidney tissue and compared these to similar networks from three mechanistically different murine models of lupus nephritis. This analysis allowed prioritization of pathogenic processes from each of the three murine models based on how well they recapitulated transcriptional processes active in human lupus nephritis. See also Hodgin *et al.* (2013) for similar analysis of murine models of diabetic nephropathy translation to glomerular transcriptional processes in human diabetic nephropathy. Cross-species analyses such as these can help understand the significance of interesting findings in preclinical model molecular networks within the human disease context.

The publication of the complete sequence of the eukaryote genome of the yeast (*Saccharomyces cerevisiae*) along with the advanced genetics techniques available and the high degree of conservation of basic processes between yeast and human cells have made yeast a powerful tool for systematic analysis of protein function and for mechanistic studies of small-molecule effects. Lum and colleagues (2004) performed a carefully controlled genome-wide screen of each of 78 compounds against pools of 3503 heterozygous deletion strains to validate previously known mechanisms of action and to identify potentially new mechanisms of action for these small molecules. The endpoint was drug-specific fitness defects of each heterozygous deletion strain in a pool of all strains compared with its average performance from a reference set. This analysis divided the compounds into three groups. The results defined one small group with no drug-specific fitness effect, likely due to the absence of the target in the screen. At the time, there were heterozygous deletion strains for only about half the yeast genes. Another small group was defined with widespread fitness changes. The last group contained the majority of the compounds with each compound causing relatively few fitness changes within the pool, indicating that the test drugs in these cases inhibit one or a small number of molecular targets within the yeast network. This last group is the most attractive for follow-up studies to confirm the mechanism of action in yeast and human cells and for potential exploitation as drug targets.

The development of global genotyping capability by array-based methods along with the development of haplotype mapping enabled genome-wide association studies (GWAS), which measure the segregation of loci (single nucleotide polymorphisms, or SNPs) across the human genome among case and control groups for diseases or complex phenotypes (e.g., height). GWAS methodology quickly matured and once study designs controlled for false discovery rate, its use expanded rapidly. To great excitement during the past decade, the results of many GWAS were published and together have provided hundreds of loci and candidate genes that strongly associate with diseases and other heritable phenotypes. Because GWAS is an unbiased and data-driven methodology, many of these associations were not expected based on our prior knowledge, providing new and interesting relationships to explore. The flip side of this is that many of these candidate genes are not connected to our prior knowledge, which limits avenues of exploration when these data are considered in the absence of any other additional experimental data.

Widespread application of global transcriptomics technologies has led to the generation of vast quantities of gene expression data that reside in publicly accessible databases like Gene Expression Omnibus (GEO), European Molecular Biology Laboratory–European Bioinformatics Institute (EMBL-EBI) and in proprietary databases. While these data have a certain value on a per-study basis, considering and curating these data as a whole is a powerful idea that has been exploited to leverage the prior investments and work of many scientists to understand the biological systems at a deeper level. We will explore a few examples of using such curated compendia of data to create tools for greater utility than the sum of the individual experiments.

Researchers at the Broad Institute have created a research tool, Connectivity Map (www.broadinstitute.org/cmap/), by performing transcriptional profiling on a number of human cell lines treated with a large panel of drugs and other compounds of interest to create a database of drug-response gene expression profiles (Lamb *et al.*, 2006; Lamb, 2007). This tool is publicly available such that anyone may upload differential gene expression profile(s) to query against the Connectivity Map. The query output is a rank-ordered list of drug-response profiles that represent testable data-driven hypotheses about the system under study. For example, submission of a differential gene expression profile generated from quiescent and activated fibroblasts would return a prioritized list of drugs whose gene expression signatures are similar, with defined probabilities, to the activated fibroblast signature. This provides a series of potential tools (drug compounds identified) useful for the dissection of the molecular mechanisms of quiescent fibroblast activation. This is limited by the number of drug-induced gene expression profiles and perhaps by the use of a limited number of cell types. Nevertheless, this is an easily accessible resource for hypothesis generation.

Sirota and colleagues (2011) have extended the idea of Connectivity Map predictions to include predictions of rank-ordered pairwise disease–drug relationships based on the differential gene expression profiles of cases and controls representing 100 diseases derived from carefully curated public gene expression data and differential expression profiles from cells treated or not with each of 164 drugs. Clustering the matrix of drug–disease relationships (164 × 100) on each axis based on similarity score calculated for each pair produced many known disease–drug clusters, but also many previously unknown relationships. The position of these novel findings within the known disease–drug clusters gives a rational pathophysiological context for interpretation of these novel findings, suggesting testable hypotheses about these novel drug–target relationships. One such observation predicted that cimetidine, an over-the-counter histamine H2-receptor antagonist used to control stomach acid production, would be potentially effective in treating lung adenocarcinoma. In their paper, they report *in vitro* and *in vivo* efficacy using a mouse xenograft model. This provides a computational means for making use of publicly available gene expression data to generate hypotheses about novel targets for disease intervention using high-quality gene expression data generated in cases and controls of the disease of interest.

Campillos and colleagues (2008) constructed a network of drug–drug relationships based on similarity of side-effects profiles and identified 1018 drug–drug relationships mediated by side-effects similarity. Many relationships were expected based on similarity of chemical structure and/or therapeutic indication; however, 261 drug–drug relationships were observed between structurally dissimilar molecules from different indications. Twenty of these were tested experimentally and 13 novel drug–target relationships were confirmed by *in vitro* binding assay, 10 with inhibition constants less than 10 nM. Nine of these were tested and confirmed in cell-based assays. Chiang and Butte (2009) review advances in computational predictions of severe adverse reactions (SADR) and adverse reactions (ADR), by highlighting several predictors of SADR and a variety of databases of both clinical and molecular measurements that are being combined computationally with additional data to yield novel insights into drug–target interactions that demonstrate efficacy and that generate SADR. Yilidrim and colleagues (2007) constructed drug–target network-based pairwise relationships between FDA-approved drugs and reported drug–target binding based on data from the DrugBank database (www.drugbank.ca; Wishart, 2006). Many drugs were interconnected in a large component of the graph with local clustering of similar drugs according to Anatomical Chemical Therapeutic classification. As expected, a significant proportion of approved drugs were "me too" drugs; however, drugs under development at the time indicated more target diversity. There were interesting topological differences between etiological and palliative drugs, which may point toward means of identifying drug targets that are more likely to ameliorate the disease condition rather than treating symptoms and may also suggest additional novel drug targets. This kind of network analysis may also indicate opportunities for poly-pharmacy – where two or more compounds with different mechanisms and points of interaction with the molecular network may more effectively moderate the disease phenotype than either drug alone.

A particularly useful approach combines biological data with drug–protein interaction data derived from high-throughput screening to begin to map drug–molecular information flow at the systems level (Fliri *et al.*, 2005a, 2005b, 2007, 2009). The core of the system is a database of interaction data between thousands of drugs and drug-like molecules and a sampling of the human proteome. Mapped to this database are high-quality protein–protein interaction (PPI) networks and drug–effects networks. Clustering of the chemical molecules by properties other than primary structure resulted in discrete groups that mapped to groups of specific indications and physiological effects. Creation of disease tissue-specific molecular networks based on the PPI and drug–protein interaction and effects allowed inference as to which parts of the network were altered by a specific drug and which altered subnetworks tend to drive effects and side effects. This system can be queried from several entry points (a desired drug effect) to follow the effects as the information is transmitted throughout the networks to the other mapped endpoints (Fliri *et al.*, 2010).

Phenotyping of knockout mice has been suggested to be a fruitful avenue of drug target identification. Zambrowski and Sands (2003) have performed a retrospective analysis of the 100 best-selling drugs and report that the phenotypes of mice with genes for the targets of these drugs knocked-out correlate well with efficacy. Several organizations have systemic mouse-knockout and phenotyping programs. Genentech executed the "Secreted Protein Discovery Initiative," a multi-pronged bioinformatics effort to identify novel secreted and transmembrane proteins (Clark *et al.*, 2003), proteins that are likely to be enriched for drug targets based on historical precedent. This effort identified 1021 genes of which 86% (879) encode for secreted or transmembrane, and 13% encode cytoplasmic or nuclear proteins; for 1%, the subcellular localization of the encoded proteins could not be determined. Of the genes identified, 25% of the cDNAs represent newly discovered proteins; 20% encode variants of existing genes; and 5% represent completely novel proteins. Genentech collaborated with Lexicon Pharmaceuticals (Woodlands, TX) to select 472 of these secreted and transmembrane proteins for construction of a library of knockout mice, which were subsequently run through a broad, unbiased phenotypic screening panel covering embryonic development, metabolism, and immune, nervous, and cardiovascular systems. Analysis of individual knockout mouse models has proven useful in many cases. Taken as a whole, such a large collection of phenotypically well-characterized knockout mice represents a large-scale systematic perturbation of the molecular networks in these mice that when integrated with other data should provide a rich picture of the states underlying many disease-relevant phenotypes. Many of these lines are available to non-profit academic and government labs through the MMRRC (Mutant Mouse Region Resource Centers supported by National Institutes of Health (NIH)). Since 2000, Pfizer has also invested significant resources to generate and extensively phenotype a large collection of knockout mice, targeting genes likely to be important for drug development in important therapeutic areas (Ozolinš, 2010; Sacca *et al.*, 2010).

Of particular relevance to target selection is global identification of changes in RNA alternate splicing and promoter preferences facilitated by carefully designed and executed RNA-Seq experiments. Genes are often taken as the unit of target, and it is become increasingly clear that alternate splicing of RNA transcripts is important in many biological processes, including disease. As these changes are more fully characterized, it is possible that isoform switching in a particular protein that occurs with the onset of disease may provide a more refined target for intervention. It is also likely that individual genetic variation at particular slice junctions may impair normal pre-mRNA processing, in some cases leading to disease. Xiong and colleagues (2015) report development of an algorithm, splicing-based analysis of variants (SPANR), to score how strongly genetic variants affect RNA splicing. They report widespread patterns of variant-driven RNA splicing dysregulation and have identified variants likely causal for mRNA missplicing with neurodevelopmental phenotypes in autism spectrum disorders.

Disease complexity – networks are fundamental

The complex diseases we seek to treat involve the interactions of many organ systems within the body, develop and present over long periods of time and are influenced by many genetic and environmental factors and the interactions between these factors. Over the past decade many genes associated with complex phenotypes and diseases have been identified. It has proven much more difficult to understand the mechanisms by which these constellations of proteins influence disease states because diseases result from the behavior of a complex dynamic adaptive system whose states are highly context-dependent and not from dysfunctions of individual gene products. In order to integrate and organize the massive scale and diversity of data now collected, network analysis techniques have been applied to and elaborated for biological systems. We have gained a very good framework understanding of complex diseases through decades of physiological, biochemical, and pharmacological research greatly extended by advances in recent decades by the molecular dissection of these processes. There are many networks operating within and between cells including gene regulatory networks, transcriptional networks, protein–protein interaction networks, and metabolite networks, and each provides a view of a facet of cellular function. These separate glimpses into cellular function are driven by our methodologies. Of course, in reality all these subcellular networks interact to form a giant network that along with individual genetic variation and environmental signals determines the state of the cell at any instant and drives the cellular behavior. Multiplying this across cell types and tissues and up to higher levels of biological organization and interactions between these various levels provides an idea of the complexity of a multi-genic disease state and its progression over time. Until recently, we have been limited by our analytical tools in our ability to reconstruct in large scale the systems that we have so finely dissected and analyzed. Many of our tools have supported textbook pictures of relatively simple biochemical and physiological processes. Now with the tools to generate vast quantities of data, we must continue to develop the techniques for accurate reconstruction of molecular networks that underlie the physiology and pathophysiology we wish to manipulate. Continuing development of network analysis techniques is allowing us to begin to examine physiology at multiple scales of integration and interaction that more accurately reflect the context within which disease-associated genes operate. We will walk though a few examples of increasingly complex network analysis in light of our goal of understanding the role of a particular candidate drug target within the context of the molecular network in the disease state of interest.

A network is a graphical representation that organizes and displays high-dimensional data where nodes in the graph represent entities (proteins, for example) and edges or lines (in our example, representing the interaction between two proteins) connecting nodes. As such, a network highlights relationships between all the nodes based on the pairwise analysis of the data used to construct the network; the structure of the network based on these relationships provides clues as to how the network behaves as a whole. One example is a co-expression network, which represents the pairwise correlation of each transcript with every other transcript expressed in the system. See Zhang and Horvath (2006) for a detailed description of weighted co-expression network analysis. Operationally, this could be presented as a topological overlap map (Figure 2.4), or a grid with the transcripts across the top and down the side, with each square in the map containing the pairwise gene expression correlation across all samples (Ravasz *et al.*, 2002). The correlation scores are typically weighted to favor the more highly correlated and therefore interconnected transcripts. Performing hierarchical clustering along each axis results in the formation of a series of discrete clusters of highly correlated transcripts along the diagonal. This topological overlap map organizes a vast amount of data and in doing so reveals, in a single visualization, the connectivity patterns of the network as a whole. The interpretation of the clusters is that transcripts within a cluster are more closely interrelated within the cluster than they are related to transcripts outside the cluster. Analysis of transcripts within a cluster or module reveals enrichment for particular biological processes or pathways reflecting the data-driven identification of biologically relevant functional modules that are related to the complex phenotype of the study subjects. Thus, co-expression analysis through topological overlap analysis identifies sets of closely regulated genes relevant to the phenotype for further investigation. Keller and colleagues (2008) extended transcript co-expression analysis to examine network connectivity in six tissues in diabetes-resistant and diabetes-susceptible strains of mice at 4 and 10

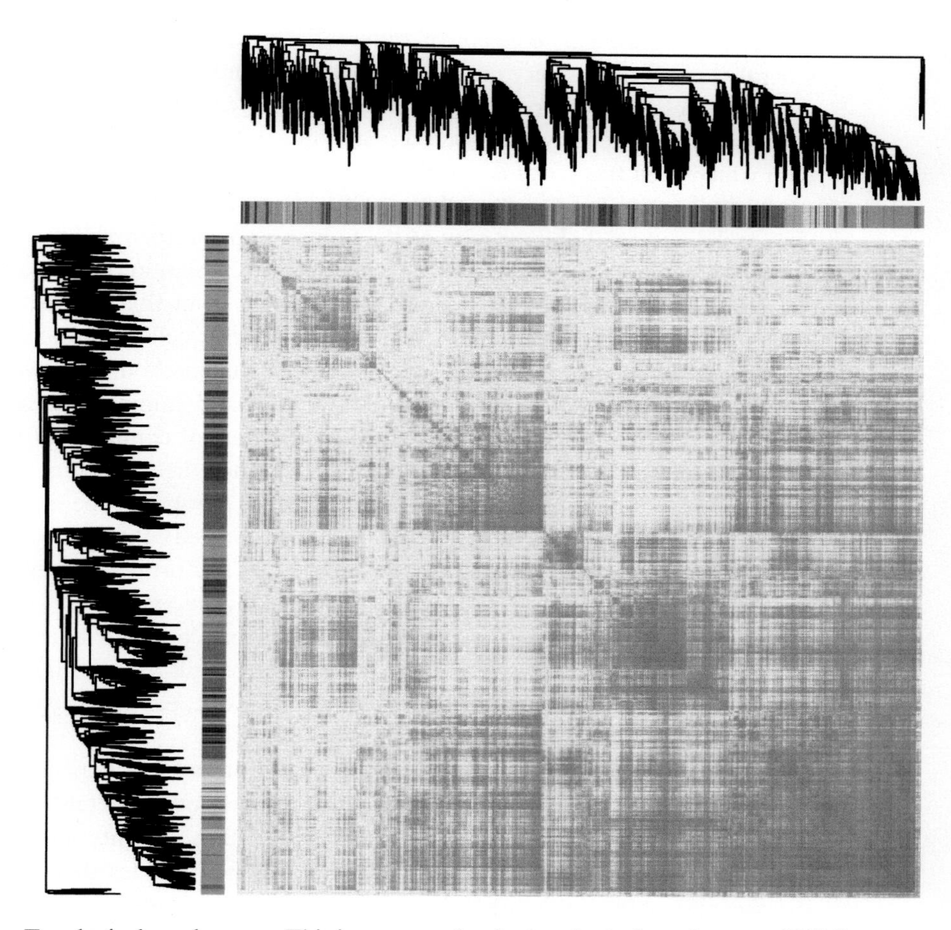

Figure 2.4 Topological overlap map. This is an example of a topological overlap map (TOM) constructed from weighted pairwise calculation of gene expression correlation between each transcript and every other transcript across the study population. Transcripts are arrayed across the x-axis and y-axis; each cell is a weighted correlation between the two corresponding transcripts. Hierarchical clustering is performed along each axis, resulting in the generation of highly interconnected modules of transcripts along the diagonal. These modules represent biologically relevant functional units related to the phenotype of the study subjects. A black and white version of this figure will appear in some formats. For the color version, please refer to the color plate section.

weeks of age, identifying 105 co-expression modules and examined the correlation structure of co-expression modules within and between these six tissues. In addition, this study highlights the importance of intercellular to intracellular networks for understanding complex disease phenotypes.

When co-expression analysis has been used to compare various phenotype groups, such as diseased and healthy tissue, different tissues or the same tissue before and after a treatment, it is observed that in different contexts transcripts within a network are interconnected in different patterns and elucidate different network modules. This concept of differential connectivity has been explored in several systems and is very important for understanding how best to target a network for molecular intervention to alter the emergent phenotype or disease state. Differential connectivity is measured as a significant change in the number of genes with which each transcript is correlated between the two phenotypic groups under study. Oldham and colleagues (2006, 2008) compared the differential gene expression and differential connectivity of the human and chimpanzee brain. It may not be surprising that on the basis of gene expression human and chimpanzee brains are fairly similar. Interestingly, modules identified from this whole-brain sample corresponded with discrete brain regions and specific cell types emphasizing the functional significance of modules identified by this method. However, there are significant functional differences between humans and chimpanzees and this may be reflected in the relatively high degree of differential connectivity between the brains of these two species. This example illustrates the important concept that while gene expression levels between two phenotype groups may be similar (and this is the most common level of analysis typically performed), the connectivity may be significantly different between these groups – this key property would have been missed if the analysis was stopped at the level of gene expression alone. Despite the similarity in gene expression levels the differential connectivity as displayed on the topological overlap maps of these two species reveals a very different modular organization of the transcriptional regulatory network, particularly at the level of the cerebral cortex compared with subcortical regions.

Wang *et al.* (2006) and Chen *et al.* (2008) performed analyses of hepatic gene expression in the F2 generation of crosses between two different inbred mouse strains on an apo-lipoprotein E null background that presents several metabolic syndrome-like traits. Along with identifying several modules that correlate with disease phenotypes, they observed differential gene expression and connectivity between the male and female mice. This is a very interesting observation and very important to understand in the context of drug discovery and target selection for drug discovery. Historically, most drug discovery and development has been performed using male animals and male research subjects. However, we are increasingly aware of major differences in risk factors and disease presentation between males and females, and the concept that there are significant differences between male and female patients in the modular structure of molecular networks underlying disease phenotypes points to the possibility that different interventions may be required to treat the same disease in male and female patients. Lum and colleagues (2006) also reported differential transcriptional connectivity in the brain of male versus female mice.

Co-expression analysis helps to organize the changes in gene expression and connectivity into a number of discrete interconnected modules of genes helping to focus analysis; however, it does not help in dividing the list of changes into those that are likely driving the phenotype of interest and those that change in response to more proximal changes. Perturbations of a system are typically used to help elucidate the cascade of changes that occur within a network that drives the phenotype of interest. Several groups have used naturally occurring genetic variation in segregating populations as a systemic, multi-factorial and relevant source of perturbations to analyze causal relationships in transcriptional networks. This analysis is based on at least two important observations. First, gene expression levels are to a significant extent heritable, such that variation in DNA at specific loci influences the level of expression of specific transcripts. Second, the fact that variations in DNA drive changes in RNA and then protein function constrains the number of possible models of relationship between DNA variation, changes in transcript levels and phenotypic measurements, simplifying and enabling the analysis.

This type of analysis combines genetic and functional genomic analysis by taking advantage of global gene expression levels as quantitative traits, global genotyping to measure variation in DNA, and phenotypic measurements. In the case of the study cited above (Chen *et al.*, 2008), hundreds of F2 offspring of the cross between two inbred mouse strains were analyzed with each F2 mouse containing a genome that was a unique mosaic of the genomes of the two parental strains representing a systematic, multi-factorial genetic variation across this F2 generation that drives changes in transcript expression and disease phenotype. By performing linkage analysis for expression and phenotypic traits, expression and phenotypic traits that are linked to the same genetic loci are identified. For these traits there is a limited number of ways in which they can be related given that they are driven by the DNA variation at that locus across the population: the expression trait can cause the phenotypic trait, the gene expression trait could be driven by the phenotypic trait, or the gene expression trait could be independent of the phenotypic trait. Causal analysis will determine which of these models best explain the data. In this manner, genes within the co-expression models can be divided into those likely to be causal for the phenotype and those likely to be responsive to the driving genes. This greatly narrows down the search for likely drug targets and points towards genes that have a higher probability of driving the disease phenotype. This type of integrated genomic analysis has been performed on many model systems and has been extended to humans (Emilsson *et al.*, 2008). Over the past 20 years, the mathematical and logical framework underlying causal analysis has been well-developed and elaborated and is widely applied in many fields. A detailed presentation is given by Pearl (2009; additional information is available at www.cs.ucla.edu/~judea).

Basso and colleagues (2005) used a novel algorithm called ARACNe (algorithm for the reconstruction of accurate cellular networks) to reverse engineer the regulatory network of human B cells using genome-wide expression profiles from a diverse set of 336 B-cell phenotypes. Using this approach, they were able to uncover with high probability the regulatory logic of this network, finding it to be scale-free and hierarchical and identifying and subsequently validating key hubs and subhubs that regulate this structure. Kidd and colleagues (2014) provided a comprehensive review of integrative biology approaches applied to immunology at a systems level. Califano and colleagues (2012) review paths forward and challenges to constructing higher levels of predictive models for understanding disease and understanding how better to target networks to modify disease.

Network-based analyses of the massive amounts of global omics data have offered many new insights into our understanding and limitations of our understanding of how to choose targets for therapeutic intervention. We are now beginning to redefine our classification of many diseases in molecular terms rather than in terms of anatomy and symptomology. For example, Perou *et al.* (2000) were able to molecularly classify breast cancer into multiple subtypes (e.g., luminal A, luminal B, basal-like). It has become apparent that there is often considerable heterogeneity in disease, and network analysis and network topological analysis (Lum *et al.*, 2013) are helping to stratify disease patients into disease subgroups based on different underlying molecular mechanisms, each likely to require a different intervention. Cancer has led the way in this regard, perhaps owing to the availability of tumor tissue for molecular analysis. A second concept that has been around for some time but usually approached empirically is that of poly-pharmacy, or treating a condition with more than one drug. The hierarchical and modular nature of molecular networks suggests that it may be difficult to modify the behavior of an entire network or subnetwork by modulating a single node with a drug. There may be certain stability or robustness within a module that may require more than one "hit" to modify its behavior. Distant modules within a hierarchical network may each require their own intervention. Building predictive models of molecular networks underlying disease will help to test such hypotheses in a systematic manner *in silico* in parallel with or in advance of laboratory validation.

Summary

There is no more exciting time to be a biologist. It is proper to view the long evolution of biology from a qualitative descriptive science to a quantitative science in the proper sense as an exponential progression. We have lifted off the asymptote and have begun an exponential climb in which the quantity and diversity of data we collect will continue to increase and the potential interaction among this growing mass of data will explode. The implications of this eventuality are manifold. First among these is that the skill set required of a biologist is transforming. We will always need to design and run experiments and to collect data, but the scale of the data collected will be enormous and the heart of the task will be the analysis and integration of these data. Therefore, emerging biologists today must be skilled in data science – meaning a mixture of hacking and computer programming skills to handle the sheer volume of data and to reduce it in a sensible manner to a form that we can process with our innate cerebral abilities. One must have a solid grasp of statistics to allow proper design and analysis of data at scale and of data arranged in large networks. Of course, one must have domain knowledge of aspects of biology and I would argue in particular at large scale – physiology, population biology and with an evolutionary perspective. It is most likely that a team of scientists with a diverse set of complementary and overlapping skills will be required to execute such analyses. The building models that reflect the complexity of the physiological systems required for future drug discovery will require new ways of working together. Such models will require iterative development and refinement that will most likely be beyond the capacity of an individual, team, or company. Data and models will need to be shared freely with sufficient documentation to allow reproduction, verification, and to allow others to build upon what has come before.

It is likely that as we gain more experience in analyzing these data and building integrated multi-scale models using these data, design motifs or underlying principles may emerge that allow us to organize data in more useful ways that simplify our understanding. I would say that one principle that may be postulated from our many failed attempts in early but particularly in late-stage drug development is that the system we study is highly constrained. By this I mean that billions of years of selection has resulted in a highly integrated and dynamic biological system that has to maintain fitness in the face of a tremendous input of environmental and internal signaling. This suggests to us that it is very easy to disrupt this balance and very difficult to achieve a particular desired outcome without any "side effects." This is so important because it constrains the space of possible solutions from something unimaginably large to a much smaller space defined by the logic of the system; our challenge is to uncover this logic. This would be analogous to unifications that occurred when Mendeleev organized the elements known in his day into the periodic table that enabled very precise predictions about then unknown elements that would soon be discovered, or the elucidation of the laws of thermodynamics that derived from the study of steam engines. Predicting the future is always a risky venture – looking to past lessons from seemingly impossible goals like landing a man on the moon and returning him safely or sequencing the entire human genome, such ventures quickly become manageable by virtual of the combined technological advancements that accrue as many dedicated minds push the limits of what is thought possible. Suffice it to say that there is much exciting work to be done in biology and in its application to drug discovery and a great deal of unmet need to be addressed. This will take the talents and creativity of many diverse groups of people to fulfill.

References

Arrowsmith, J. Trial watch: Phase III and submission failures: 2007–2010. *Nature Reviews Drug Discovery*. 2011a;10:87.

Arrowsmith, J. Trial watch: Phase II failures: 2008–2010. *Nature Reviews Drug Discovery*. 2011b;10:328–329.

Basso, K., Margolin, A. A., Stolovitzky, G., *et al.* Reverse engineering of regulatory networks in human B cells. *Nature Genetics*. 2005;37:382–390.

Berthier, C. C., Bethunaickan, R., Gonzalez-Rivera, T., *et al.* Cross-species transcriptional network analysis defines shared inflammatory responses in murine and human lupus nephritis. *Journal of Immunology*. 2012;189:988–1001.

Bleicher, K. H., Böhm, H.-J, Müller, K. and Alanine, A. I. Hit and lead generation: Beyond high-throughput screening. *Nature Reviews Drug Discovery*. 2003;2:369–378.

Califano, A., Butte, A. J., Friend, S., Ideker, T. and Schadt, E. Leveraging models of cell regulation and GWAS data in integrative network-based association studies. *Nature Genetics*. 2012;44: 841–847.

Campillos, M., Kuhn, M., Gavin, A.-C., Jensen, L. J. and Bork, P. Drug target identification using side-effect similarity. *Science*. 2008;321:263–266.

Chen, Y., Zhu, J., Lum, P. Y., *et al.* Variations in DNA elucidate molecular networks that cause disease. *Nature*. 2008;452:429–435.

Chiang, A. P. and Butte, A. J. Data-driven methods to discover molecular determinants of serious adverse drug events. *Clinical and Pharmacological Therapeutics*. 2009;85:259–268.

Clark, H. F., Gurney, A. L., Abaya, E., *et al.* The secreted protein discovery iniative (SPDI), a large-scale effort to identify novel human secreted and transmembrane proteins: A bioinformatics approach. *Genome Research*. 2003;13:2265–2270.

Drews, J. Drug discovery: A historical perspective. *Science* 2000;287:1960–1964.

Drews, J. and Ryser, S. The role of innovation in drug development. *Nature Biotechnology*. 1997;15:1318–1319.

Emilsson, V., Thorleifsson, G., Zhang, B., *et al.* Genetics of gene expression and its effects on disease. *Nature*. 2008;452:423–430.

Edwards, A., Isserlin, R., Bader, G. D., *et al.* Too many roads not taken. *Nature*. 2011;470:163–165.

Fliri, A. J., Loging, W. T., Thadeio, P. F. and Volkmann, R. A. Biospectra analysis: Model proteome characterizations for linking molecular structure and biological response. *Journal of Medicinal Chemistry*. 2005a;48:6918–6925.

Fliri, A. J., Loging, W. T., Thadeio, P. F. and Volkmann, R. A. Analysis of drug-induced effect patterns to link structure and side effects of medicines. *Nature Chemical Biology*. 2005b;1:389–397.

Fliri, A. J., Loging, W. T. and Volkmann, R. A. Analysis of system structure–function relationships. *ChemMedChem*. 2007;2:1774–1782.

Fliri, A. J., Loging, W. T. and Volkmann, R. A. Drug effects viewed from a signal transduction network perspective. *Journal of Medicinal Chemistry*. 2009;52:8038–8046.

Fliri, A. J., Loging, W. T. and Volkmann, R. A. Cause–effect relationships in medicine: A protein network perspective. *Trends in Pharmaceutical Science*. 2010;31:547–555.

Forrest, M. J., Bloomfield, D., Briscoe, R. J., *et al.* Torcetrapib-induced blood pressure elevation is independent of CETP inhibition and is accompanied by increased circulating levels of aldosterone. *British Journal of Pharmacology*. 2008;154:1465–1473.

Hodgin, J. B., Nair, V., Zhang, H., *et al.* Identification of cross-species shared transcriptional networks of diabetic nephropathy in human and mouse glomeruli. *Diabetes*. 2013;62:299–308.

Inazu, A., Brown, M. L., Hesler, C. B., *et al.* Increased high-density lipoprotein levels caused by common cholesterol-ester transfer protein gene mutation. *New England Journal of Medicine*. 1990;323:1234–1238.

Keller, M. P., Choi, Y., Wang, P., *et al.* A gene expression network model of type 2 diabetes links cell cycle regulation in islet with diabetes susceptibility. *Genome Research*. 2008;18:706–716.

Kidd, B. A., Peters, L. A., Schadt, E. E. and Dudley, J. T. Unifying immunology with informatics and multiscale biology. *Nature Immunology*. 2014;15:118–127.

Krejsa, C. M., Horvath, D., Rogalski, S. L., *et al.* Predicting ADME properties and side effects: The bioprint approach. *Current Opinion in Drug Discovery and Development*. 2003;6:470–480.

Lamb, J. The connectivity map: A new tool for biomedical research. *Nature Reviews Cancer*. 2007;7:54–60.

Lamb, J., Crawford, E. D., Peck, D., *et al.* The connectivity map: Using gene-expression signatures to connect small molecules, genes and disease. *Science*. 2006;313:1929–1935.

Loging, W., Rodriguez-Estaban, R., Hill, J., Freeman, T. and Miglietta, J. Cheminformatic/bioinformatic analysis of large corporate databases: Application to drug repurposing. *Drug Discovery Today: Therapeutic Strategies*. 2011;8:109–116.

López-Muñoz, F., Alamo, C., Cuenca, E., *et al.* History of the discovery and clinical introduction of chlorpromazine. *Annals of Clinical Psychiatry*. 2005;17:113–135.

Lum, P. Y., Armour, C. D., Stepaniants, S. B., *et al.* Discovering modes of action for therapeutic compounds using a genome-wide screen of yeast heterozygotes. *Cell*. 2004;116:121–137.

Lum, P. Y., Chen, Y., Zhu, J., *et al.* Elucidating the murine brain transcriptional network in a segregating mouse population to identify core functional modules for obesity and diabetes. *Journal of Neurochemistry*. 2006;97(Suppl 1):50–62.

Lum, P. Y., Singh, G., Lehman, A., *et al.* Extracting insights from the shape of complex data using topology. *Scientific Reports*. 2013;3:1236.

Morabia, A., Cayanis, E., Costanza, M. C., *et al.* Association of extreme blood lipid profile phenotypic variation with 11 reverse cholesterol transport genes and 10 non-genetic cardiovascular disease risk factors. *Human Molecular Genetics*. 2003;12:2733–2743.

Munos, B. Lessons from 60 years of pharmaceutical innovation. *Nature Reviews Drug Discovery*. 2009;8:959–968.

Oldham, M. C., Horvath, S. and Geschwind, D. H. Conservation and evolution of gene coexpression networks in human and chimpanzee brain. *Proceedings of the National Academy of Sciences*, USA. 2006;103:17973–17978.

Oldham, M., Konopka, G., Iwamota, K., *et al.* Functional organization of the transcriptome in human brain. *Nature Neuroscience*. 2008;11:1271–1282.

Overington, J. P., Al-Lazikani, B. and Hopkins, A. L. How many drug targets are there? *Nature Reviews Drug Discovery*. 2006;5:993–995.

Ozolinš, T. R. S. De-risking developmental toxicity-mediated drug attrition in the pharmaceutical industry. In *Predictive Toxicology in Drug Safety*, ed. J. J. Xu and L. Urban (pp. 153–182). Cambridge: Cambridge University Press, 2010.

Palmer, A. M. and Stephenson, F. A. CNS drug discovery: Challenges and solutions. *Drug News Perspectives*. 2005;18:51–57.

Pearl, J. *Causality: Models, Reasoning and Inference* (2nd edn). Cambridge: Cambridge University Press, 2009.

Perou, C. M., Sorlie, T., Eisen, M. B., *et al.* Molecular portraits of human breast tumours. *Nature*. 2000;406:747–752.

PhRMA. Profile of the Biopharmaceutical Research Industry (2014). www.phrma.org/sites/default/files/pdf/2014_PhRMA_Profiles.pdf.

Preskorn, S. H. CNS drug development: Part 1: The early period of CNS drugs. *Journal of Psychiatric Practice*. 2010a;16:334–339.

Preskorn, S. H. CNS drug development: Part 2: Advances from the 1960s to the 1990s. *Journal of Psychiatric Practice*. 2010b;16:413–415.

Ravasz, E., Somera, A. L., Mongru, D. A., Oltvai, Z. N. and Barabasi, A.-L. Hierarchical organization of modularity in metabolic networks. *Science*. 2002;297:1551–1555.

Saboo, A. Biopharma posts a chart-topping 41 new drug approvals in 2014. FierceBiotech (www.fiercebiotech.com). 2015, Jan 2.

Sacca, R., Engle, S. J., Qin, W., *et al.* Genetically engineered mouse models in drug discovery research. *Methods in Molecular Biology*. 2010;602:37–54.

Schadt, E. E., Friend, S. H. and Shaywitz, D. A. A network view of disease and compound screening. *Nature Reviews Drug Discovery*. 2009;8:286–295.

Sirota, M., Dudley, J. T., Kim, J., *et al.* Discovery and preclinical validation of drug indications using compendia of public gene expression data. *Science Translational Medicine*. 2011;3:1–10.

Swinney, D. C. and Anthony, J. How were new medicines discovered? *Nature Reviews Drug Discovery*. 2011;10:507–519.

Tall, A. R. CETP inhibitors to increase HDL cholesterol levels. *New England Journal of Medicine*. 2007;356:1364–1366.

Tian, J. and Patel, J. M. TALE: A tool for approximate large graph matching. In *Proceedings of the 24th International Conference on Data Engineering, Cancun, Mexico* (pp. 963–972). Washington, DC: IEEE Computer Society, 2008.

Wang, S., Yehya, N., Schadt, E. E., *et al.* Genetic and genomic analysis of a fat mass trait with complex inheritance reveals marked sex specificity. *PLoS Genetics*. 2006;2: e15.

Wishart, D. S. Drugbank: A comprehensive resource for *in silico* drug discovery and exploration. *Nucleic Acids Research*. 2006;34: D668–D672.

Xiong, H. Y., Alipanahi, B., Lee, L. J., *et al.* The human splicing code reveals new insights into the genetic determinants of disease. *Science*. 2015;347:1254806.

Yildirim, M. A., Goh, K.-I., Cusick, M. E., Barabasi, A.-L. and Vidal, M. Drug–target network. *Nature Biotechnology*. 2007;25:1119–1126.

Zambrowski, B. P. and Sands, A. T. Knockouts model the 100 best-selling drugs – Will they model the next 100? *Nature Reviews Drug Discovery*. 2003;2:38–51.

Zhang, B. and Horvath, S. A general framework for weighted gene co-expression network analysis. *Statistical Applications in Genetics and Molecular Biology*. 2005;4: Article 17.

3 Understanding human disease knowledge through text mining: What is text mining?

Raul Rodriguez-Esteban

The aim of text mining in biomedicine is to extract valuable information from large amounts of biomedical text. For this purpose it borrows techniques from fields such as natural language processing (NLP), information retrieval (IR), information extraction (IE), and artificial intelligence (AI). However, many of these techniques need to be adapted to the particularities of biomedical text, because this text possesses a unique diversity of vocabularies and writing styles, as can be seen in clinical narratives, regulatory reports, and scientific articles. For example, an NLP algorithm that recognized sentences in newspapers would need to be adjusted for biomedical text, because periods that do not separate sentences are used more frequently in biomedical text than in newspapers, which would disorient the NLP algorithm (Tomanek *et al.*, 2007). The particular information needs in biomedicine have also led to the development of specialized text-mining techniques for extracting knowledge specific to the biomedical domain, such as, for example, molecular events, perturbations and interactions.

Pharmaceutical companies are data-intensive organizations whose success depends on their ability to efficiently process large quantities of data from internal and external sources. Much valuable knowledge is locked within textual sources such as patents, clinical records, conference abstracts, and full-text articles. The growth of these textual sources means that even experts on a subject matter cannot cope with the content appearing in their niche. For example, more than 27,000 articles mentioning diabetes were listed in PubMed during the year 2013. Text mining enables the processing of such documents within practical time frames and impacting every stage of the drug discovery pipeline.

Before the late 1990s, IR was the main research field that dealt with biomedical documents. Its main focus was on improving access to literature records from biomedical databases such as Medline, a comprehensive database of scientific abstracts managed by the US National Library of Medicine (NLM). Then, in 1996, the launch of PubMed made available the majority of Medline content online (Canese, 2006). This event was followed by an increase in research about biomedical documents with a scope broader than IR. Such research was coined "text mining" due to

Bioinformatics and Computational Biology in Drug Discovery and Development, ed. W.T. Loging. Published by Cambridge University Press.

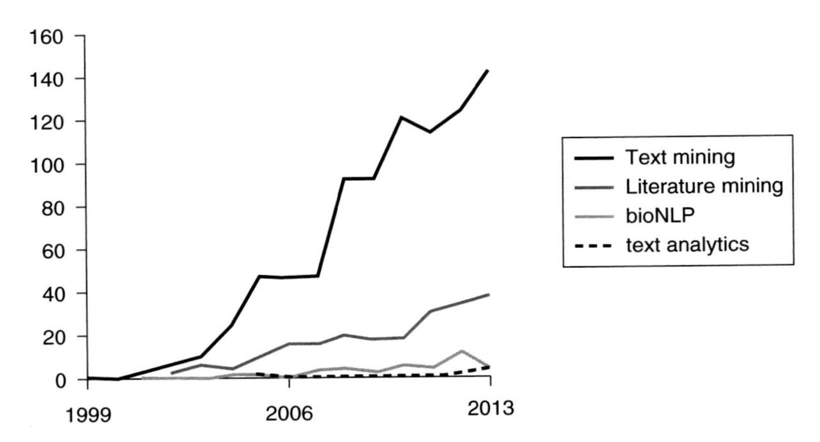

Figure 3.1 Mentions of text mining, literature mining, bioNLP and text analytics in Medline abstracts.

the emergence of data and text mining during the same period (Rodriguez-Esteban, 2008). The first publication dealing with biomedical text that used the name "text mining" came from the National Institutes of Health (NIH) in 1999 (Tanabe *et al.*, 1999). The label "literature mining" has been sometimes used to refer specifically to the text mining of the scientific literature or *bibliome* (Searls, 2001).

Names for technical and scientific fields can evolve over time. Text mining for biomedicine owes less to the fields of text and data mining, which are often concerned with identifying patterns rather than facts, than to the fields of NLP, IR, and IE. Therefore, some have preferred the name BioNLP to the name text mining. More recently, the name "text analytics" has been gaining currency. Although text mining and text analytics span many similar applications, the name text analytics is not yet popular in the biomedical realm. A search in PubMed evidences the popularity of the different labels as of 2013: text mining > literature mining > BioNLP > text analytics (Figure 3.1). The label "text mining" is implied to mean "biomedical text mining." This is a simplification that is conventional in the field and that is used throughout this chapter as well.

Sources of content

The text mining process always starts with the gathering of a corpus. In text mining, "corpus" is a fancy Latin word that simply means a group of documents. Documents may come from many sources of different quality, cost, size, and complexity. Assembling the right corpus when starting a text-mining exercise is similar to a miner choosing the right place to start digging. In other words, the characteristics of the corpus greatly affect the results of text mining.

Due to the nature of scientific research, facts are often repeated across documents and presented with different wording and in different contexts. For example, the fact that the gene *p53* suppresses the formation of tumors is mentioned in a multitude of Medline records but worded in many different ways, e.g., "*p53* efficiently suppresses tumor development" and "regulation of the p53 tumor suppressor protein." The more variety in the wording used in the corpus the higher the chances it will be detected at least once. For example, a text-mining algorithm only acquainted with British English would still be able to find articles mentioning *p53* "tumor" suppression in Medline.

Identifying a fact once is enough for the purposes of many pharmaceutical questions (Divoli, 2006; Gomes *et al.*, 2006; McIntosh and Curran, 2009). A thorough mining effort, on the other hand, may aim to capture every single mention of a fact, even for little-known facts that might be buried in unheralded publications, purposefully concealed in patents or briefly exposed in conference abstracts. These little-known facts are like "rare cases" in data mining (Weiss, 2004) and can be highly valuable because few might be aware of their existence and thus represent a competitive advantage for a pharmaceutical company. One obvious way to increase the chances of finding little-known facts is to process as much content as possible. From this point of view, "size matters" for a corpus. However, some sources of text may contain very little relevant information and bring processing costs and noise that affect performance. Thus, the first step in a text-mining exercise is not gathering as much content as possible, as in the "size matters" approach, but gathering as much *relevant* content as possible.

Much biomedical content is not freely available and requires expensive licensing fees. Many publishers of scientific literature, for example, request fees to allow the text mining of their content even for users who already have reading access (Van Noorden, 2012, 2013, 2014). Thus, the availability of content is a key limitation for creating a comprehensive corpus and only companies with a certain budget can afford extensive access. This limitation is an important challenge in pharmaceutical settings because a text-mining exercise may miss relevant results that do appear in internet search engines such as Google and Google Scholar, which are allowed to index the content behind many paywalls, achieving a reach superior to many scientific search engines (see for example Shultz, 2007; Kulkarni *et al.*, 2009). Competing with Google's recall is a standard that text miners need to meet to the extent technically possible.

The following sections describe important sources of content for creating corpora.

Scientific abstracts and articles

Scientific publications are a primary source of knowledge in the pharmaceutical industry because they lay the foundation for a large percentage of drug discovery projects (Swinney and Anthony, 2011; Eder *et al.*, 2014). Scientific publications have also been an important focus of text-mining research. Such research has been mostly oriented towards abstracts from Medline, because Medline is a large, up-to-date and freely available resource. However, Medline has some limitations worth taking into consideration. Medline's selection of journals is curated following a set of guidelines which do not necessarily coincide with the information needs of pharmaceutical companies. Abstracts from non-US and non-English journals are relatively under-represented in Medline (Primo Peña *et al.*, 2004). Medline also has limited coverage of conference abstracts. Abstracts from many scientific conferences are only accessible to conference attendees or through payment and they are released in various formats, from HTML web pages to PDF files stored in DVD. This variability means that tailored pre-processing steps are required prior to analyzing them.

Several commercial abstract databases are available that include content beyond Medline's coverage, such as Embase, Biosis Previews, Medmeme, Chemical Abstracts and Derwent Drug File. Each of these databases has its own strengths and weaknesses. Medmeme specializes in conference abstracts. Biosis Previews has a focus on biological sciences. Embase is similar to Medline in breadth, but has different inclusion criteria for journals, so that it has more European and less American content than Medline. Derwent Drug File covers mainly drug-related journals and provides summaries of publications that are written by curators. An important problem when integrating some of these abstract databases is the large fraction of overlapping content, which requires the elimination of duplicated content (deduplication; Jiang *et al.*, 2014).

Despite the size of existing abstract databases, these databases only cover a fraction of the knowledge published in scientific publications. Abstracts of scientific articles only contain about 7% of the facts reported in the articles (Van Landeghem *et al.*, 2012). Thus, full-text articles are invaluable for text mining. However, accessing full-text articles is, at the moment, fraught with problems. The main repositories of full-text articles are PubMed Central and Europe PubMed Central (The Europe PMC Consortium, 2014), which centralize the archival of publications to support open-access (OA) mandates, author self-archival and OA journals, such as the BioMed Central and PLoS family of journals. Only the OA Subset of PubMed Central, however, is available for text mining.

Many major funding agencies and universities have issued mandates requesting that scientific publications produced with their support become OA. The ROARMAP database provides a comprehensive list of those mandates. Unfortunately, those mandates are insufficiently followed, with the result that only 20% of all scientific articles were available as OA in 2008 (Björk *et al.*, 2010). Moreover, many OA articles have licenses that do not allow text mining, or their licensing status is unclear. Establishing the licensing rights of each individual scientific article from each of the myriads of publishers is a daunting task. Pharmaceutical companies address this issue by setting agreements with intermediary institutions and companies that specialize in managing licensing rights such as ReprintsDesk, Infotrieve and Copyright Clearance Center. Such agreements enable the creation of full-text corpora for text mining, which can be created with tools such as the link-resolvers Quosa and Pubget.

Link-resolving tools are able to download articles in bulk by accessing publisher websites at speeds that do not overburden the publishers' servers. The full-text documents delivered by link-resolving tools come in XML, HTML, and PDF format, depending on the availability of formats offered by each publisher. Before text mining, these documents need to be converted, typically, to a uniform XML format, which can lead to a loss in the quality of the content, especially when the original documents are in PDF format. PDF conversion to XML poses multiple challenges, among them: (1) identifying the page layout and elements such as columns, figures, and tables; (2) resolving word wrappings and paragraph boundaries; and (3) translating special characters. For "non-natively digital" PDF documents (i.e., PDF documents based on images), optical character recognition (OCR) processing is necessary. Tools specifically built for the "interpretation" of scientific articles in PDF format have been developed, such as PDFX (Constantin *et al.*, 2013) and LA-PDFText (Ramakrishnan *et al.*, 2012). Fortunately, recent publications are increasingly more likely to be available in HTML and XML format, but the legacy PDF-only literature still needs to be taken into account.

Link-resolving tools are slow, but no firm alternatives exist yet. There are currently several initiatives by content brokers such as Elsevier, Infotrieve, and ReprintsDesk to develop application programming interfaces (APIs) that allow some form of full-text search across journals from multiple publishers and bulk download in XML format. To compete with link-resolving tools, such APIs will need access to a large share of journals and will still face the problem of integrating document formats used by different publishers. Ideally, these APIs should also have access to images and supplementary data sets. Technological hurdles are not the obstacle for such a goal, and workable business models will hopefully lead to progress in this area. At the moment, limitations in the access to full-text content for text mining still hinder the drug discovery process.

The resolution of these issues is being fostered by the Pharma Documentation Ring (P-D-R) consortium, which brings together large pharmaceutical companies interested in the discussion and advancement of topics in pharmaceutical information science (Clark *et al.*, 2013). The P-D-R has proposed that journal publishers adopt a common XML format for their articles, such as the Journal Article Tag Suite (JATS) designed by the NLM and the National Information Standards Organization (NISO). The JATS standard is currently used in PubMed Central and allows the representation of article elements such as sections, tables, figures, equations, and cross-references. However, JATS is not a standard designed specifically for text mining. For example, article section names (e.g., introduction, methods) are not semantically defined in JATS, which has led to a large multiplicity of section names being used in PubMed Central (Martin Romacker, P-D-R Special Meeting 2013). An XML standard for text mining would include, for example, tagging standards for annotations and sentence boundaries, unique identifiers for cross-references to databases and other features. That said, the adoption of JATS or any other similar XML standard by the publishing industry would be a great improvement over the present state of affairs.

Patents

Patents are instrumental for the protection of intellectual property in drug discovery. Beyond a legal document, they are also a window into competitors' activities, technologies and experimental results. Interestingly, scientific papers rarely reference patents (Schmoch, 1997) and, thus, biological data exclusively published in patents may go unacknowledged by academic researchers and absent from the scientific literature. This, however, does not diminish their usefulness for drug discovery. The pursual of "patent mining" can yield important facts of direct relevance for drug discovery. Their relative obscurity may make them even more valuable.

Currently, the largest body of freely available biomedical text is, in fact, within the set of patents of the United States Patent and Trademark Office (USPTO), which is accessible by bulk download from Google Patents. Notably, these patents are complete, with figure images, supplementary files (e.g., for DNA and amino acid sequences) and Complex Work Units, which describe patent elements such as chemical drawings, mathematical formulae, and protein structures. The unique richness of this resource allows the combination of text mining with other types of analyses such as figure mining (Rodriguez-Esteban and Iossifov, 2009) and sequence analysis. One of the drawbacks of USPTO patents, however, is that the XML formatting and data type definitions (DTD) have changed over the years.

Other major collections of patents such as European, World, Japanese, and Chinese patents are not available for free as high-throughput bulk download. Accessing these patents in bulk requires paying fees to each patent agency and dealing with different XML formats and update schedules. Alternatively, resellers such as IBM, Patbase and Thomson Reuters offer up-to-date patent collections with consistent XML format as well as additional features such as enhanced search interfaces, chemical annotations, language translation, patent family clustering, and patent landscaping.

Clinical trials

Clinical trial records are an important source of information for clinical trial designs, clinical biomarkers, competitive information, and drug safety. The only major regulatory agency providing bulk download of clinical trials is the US NIH through ClinicalTrials.gov, which covers clinical trials in the US and many performed in other countries (as of 2014, 44% of trials registered in ClinicalTrials.gov were non-US). Other sources of trial information such as the Japanese public clinical trial database UMIN, the World Health Organization's ICTRP, the International Standard Randomised Controlled Trial Number's registry and the European Medicines Agency's EudraCT provide limited download capabilities. The Cochrane Library, a valuable database of clinical trial abstracts, is not available for free bulk download.

A major drawback of most clinical trial databases is their lack of information about the outcomes of trials. US legislation requires publication of trial outcomes in ClinicalTrials.gov, but compliance is very low (Jones *et al.*, 2013). Some outcome information can be found in scientific publications, but their presentation can be incomplete and expressed differently in each publication. Moreover, many clinical trial results are never published. Even in the case of large clinical studies (more than 500 patients), slightly over 20% are not published after 6 years, possibly to hide inconvenient results or to protect intellectual property (Jones *et al.*, 2013). There are some initiatives currently to request more detailed access to clinical trial outcomes; however, regulatory agencies and pharmaceutical companies are concerned that there could be loss of privacy for clinical trial participants.

Companies such as Informa, Springer, and Thomson Reuters provide clinical trial information databases integrating many different sources, including media releases and pipeline information from company websites. In some cases they even provide detailed information that can be used for trial design and drug pharmacokinetic modeling. These databases can be purchased in XML format and mix curated annotations and free-text components that can be mined.

Grants

Academic research can long pre-date publication. Researchers move on to new areas even before publications of past work are released. Thus, scientific publications are a window into past work. Grant proposals, on the other hand, offer a view into future scientific developments and thus an opportunity to tap into emerging scientific research by establishing collaborations, partnerships and technology transfer agreements. Grant information can also be used to identify key opinion leaders and experts in areas of interest, as well as to pinpoint "hot," novel, or more established research areas.

The best resource for grant information available for text mining is the NIH Reporter (see also grants.gov), which can be downloaded in bulk. Other agencies offer more limited access. The Grist API from Europe PMC is the latest improvement in the area of grant information access as it allows queries for grants funded by a number of, mainly European, agencies (The Europe PMC Consortium, 2014). Medline also offers funding information for some publications. This information is represented in Medline XML inside the <GrantList> element.

Curated text and annotations

Curated text is free text that has been written with the purpose of summarizing or highlighting important facts and references from the primary literature. Curated text is easier to mine than other types of text because it tends to have a better-defined focus. Many biological databases contain curated text in the form of free-text annotations, such as the Gene Expression Omnibus, UniProt and Pfam. NCBI's Gene database (formerly known as Entrez Gene) includes the Gene Reference Into

Function (GeneRIF) annotations, which are functional annotations about genes written by scientific authors. Online Mendelian Inheritance in Man (OMIM) provides in-depth summaries of research about diseases. However, OMIM requires licensing for its use in commercial settings. As has been mentioned, Derwent Drug File is a commercial database that offers summaries of scientific abstracts focused on drugs.

Annotations that are not made of free text can also be mined. Typically these annotations are based on ontologies (ontologies are discussed further later), such as the gene annotations recorded by the Gene Ontology (GO). Notably, Medline is annotated by NLM curators with substance names and the Medical Subject Headings (MeSH) ontology, which covers a broad range of biomedical topics. MeSH annotations in Medline are modified by qualifiers that add nuance and expressivity. Another ontology for annotation similar to MeSH is Emtree, which is a commercial ontology used for annotation in the Embase abstract database. Emtree is similar in scope to MeSH but larger in size and without qualifiers.

Safety

Drug-related safety and toxicity analysis and monitoring have been gaining importance. Safety and toxicity information is used and produced at each step of the pharmaceutical pipeline and therefore can be found in preclinical, clinical, and post-marketing studies, as well as in other sources (Golder and Loke, 2012). Some of these sources have been specifically created for safety monitoring and are available for download, such as the FDA drug labels, the FDA Adverse Event Reporting System (FAERS), the Side Effect Resource (SIDER) from the European Molecular Biology Laboratory (EMBL), NLM's TOXNET databases, and the commercial Adis Reactions Database. Some other sources of information are collected by companies directly from patients and care providers. One type covers so-called spontaneous reports, such as phone calls and e-mail messages by patients and healthcare providers regarding marketed drugs. The increase in communication by patients through new technologies such as social media and mobile devices is opening a new frontier in safety monitoring that has yet to be addressed.

Text mining can play several roles in safety and toxicity, especially at the preclinical stage. The very slowly changing regulatory landscape, however, hinders further developments at the clinical and post-marketing stages because text mining has not been fully embraced by regulatory agencies. Interest in text mining was signaled in 2012 by the Center for Drug Evaluation and Research (CDER), a division of the US Food and Drug Administration (FDA), when it announced that it was performing a trial of the text mining tool I2E from the company Linguamatics. One of the main potential applications of text mining for safety monitoring is that of "signal detection," which involves recognizing potential connections between drugs and adverse events.

Websites

Biotechnology and pharmaceutical company websites are sources of pipeline information and licensing opportunities. They can be valuable to identify small companies that are developing new technologies or certain targets. Such companies can have limited funding and marketing capabilities but still have a website that describes their work. These company websites can be explored by web search APIs such as Yahoo!'s or by customized web crawlers that can use a list of company URLs.

News

News reports are at the cutting edge of competitive intelligence, providing updates on drug pipeline events before these are recorded in pipeline databases. Thus, news can offer the first available mentions of certain terms, such as new drug names. Monitoring and analyzing news reports, moreover, helps the understanding of pharmaceutical industry trends such as strategic investments in disease areas, company acquisitions, mergers, and licensing.

Some news sources can be accessed directly from RSS feeds. Other news freely available online are not in such convenient format but can be web-scraped and pre-processed (Breiner and Rodriguez-Esteban, 2013). A number of commercial providers such as Thomson Reuters offer collections of news centered around the pharmaceutical sector.

Internal company documents

Internal company documents such as laboratory notebooks, clinical trial reports, and regulatory submissions are difficult to access even within pharmaceutical companies due to confidentiality concerns. Laboratory notebooks, for example, contain valuable information spanning years of experimentation. However, while some companies have made an effort to digitize old notebooks and to introduce electronic notebooks, the main motivation for electronic record-keeping has been intellectual property protection rather than exploiting the trove of information that they represent. Regretfully, thus, it is generally easier to acquire external sources of information than to access internal company documents. One way to persuade stakeholders about the utility of text mining this information is by proposing initiatives oriented towards specific goals such as repositioning old drugs (Loging *et al.*, 2011).

Old internal company documents can be challenging to analyze as they may have been scanned and processed with OCR software (for an example of analysis of internal clinical trial documents, see Matthew E. Crawford's presentation "Text Mining in the Clinical Sphere: Coping with Really Unstructured Text" at the Linguamatics Users Conference, 2011). Another important resource scattered through corporate electronic data warehouses is slide presentations in PowerPoint format. These presentations can provide detail about each drug discovery project as they advance through the drug pipeline. They can be found in content management systems such as Sharepoint and Documentum, or in local hard drives. This resource was called the PowerPointome by the author and colleagues while working at Pfizer in 2008.

Social media

The explosion of social media platforms such as blogs, microblogs, discussion forums, and social networks has opened opportunities to better understand the needs of patients and healthcare providers. For example, it is now possible to track the response of key opinion leaders in the medical field after the announcement of new clinical trial data through sentiment analysis of Twitter feeds. Patients also express opinions about their condition and the drugs they take through social media. One drawback of social media content is that users may not disclose key information such as, for example, comorbidity and demographic information, that puts their statements into context.

Obtaining content from social media is not straightforward. Twitter is one of the most accessible resources, but it does not yet make available a long-term archive. Blogs and opinion forums have to be mined individually or are managed by social media companies that offer access at a cost. A mining exercise of such diverse sources can be easier to do if the focus is narrowed to an area of manageable size (Leaman *et al.*, 2010; Xu *et al.*, 2014). There are companies specialized in mining multiple social media that offer services such as analysis of trends and translation of foreign language content.

A pharmaceutical company needs to be careful when mining social media because any adverse event reported that is linked to the company's drugs needs to be communicated to regulatory authorities (a workable solution was shown at the presentation "Text Analytics in Big Pharma; Challenges and Use Cases" by Rob Hernandez from AstraZeneca at the Linguamatics Spring Users Conference, 2013). Guidances on social media usage and monitoring by pharmaceutical companies remain evolving and under discussion. It is possible that pharmaceutical companies will be required in the future to monitor social media and text mining could play a role in this arena.

Clinical records

Clinical records are the focus of important text-mining research in academia, but privacy concerns represent a hurdle for their use in pharmaceutical drug discovery. Studies involving clinical records need to be approved by research boards and are limited in scope. Thus, the adoption of electronic medical records in many countries does not yet represent an important opportunity for text mining by pharmaceutical companies except in cases in which hospitals and other institutions play an intermediary role, such as in the Framingham Heart Study and the Gutenberg Health Study.

Clinical records are challenging to mine due to the diversity, idiosyncrasies, and haphazard nature of the writing. Clinicians have little time to write notes and use abbreviations that might be particular to a hospital or clinic. Many clinical notes are simple sentences or phrases that use technical terms. For example, clinicians may use negations profusely to indicate the results of multiple negative tests. Moreover, clinical records are written in local languages and there is a lack of non-English text-mining tools and terminological resources, which are costly to create (see, for example, Eriksson *et al.*, 2013). Notwithstanding, over the last few years there has been a large increase in text-mining research of clinical records.

Other

Other types of documents can yield interesting information for drug discovery, such as content from knowledge databases provided by vendors, medical practice guidelines and books (see NLM's Bookshelf). In particular, vendor knowledge databases can have large free-text descriptions written by curators that can be explored through text mining. An example is the competitive intelligence database Thomson Integrity.

General considerations about content

Timeliness

Managing the risk profile of a drug portfolio is crucial for a pharmaceutical company and text mining needs to adapt to such needs, whether these come from drug projects involving novel mechanisms of action for so-called "first-in-class" drugs or from drug projects involving proven mechanisms of action for so-called "follow-on," "me-too," or "best-in-class" drugs. Old research findings lose relevance over the years for pharmaceutical companies due to changes in disease paradigms and the assumption that old therapeutic concepts have already been explored by competitors. On the other hand, novel targets, which have limited supporting evidence, may fit better with the latest understanding of a disease. Thus, to stay ahead of competitors, pharmaceutical companies place large value on quickly identifying recently published documents such as relevant patents and critical clinical trials, or new articles that describe promising targets. Text-mining strategies need to be able to utilize the latest content and be re-scheduled periodically to produce up-to-date results.

Diversity

Content diversity represents an important challenge for text mining. The variation in writing styles across sources is a challenge for computational approaches that are finely tuned to identify certain textual patterns in order to maximize performance (Friedman *et al.*, 2002; Caporaso *et al.*, 2008). Problems of adaptation to different types of text have been described, for example, in algorithms that recognize protein names (Kabiljo *et al.*, 2009), interactions (Tikk *et al.*, 2010), and sentences

(Clegg and Shepherd, 2007). Even within a single biomedical document, differences across document sections can be important. For example, many text-mining algorithms have been trained on article abstracts, which are quite different from the article's full text (for further discussion, see Cohen *et al.*, 2010; Martin *et al.*, 2004; Schuemie *et al.*, 2004). Thus, performance is reduced when algorithms trained on abstracts are used for full text (Verspoor *et al.*, 2012).

To address this problem, algorithms can be re-trained for every type of text or trained over several types of text. However, this involves a considerable effort. Another option is to create an algorithm with the ability to adapt to the characteristics of any given text. Such an approach is called "domain adaptation." Domain-adaptation strategies have been proposed for a number of tasks (examples include: Pyysalo *et al.*, 2006; Dahlmeier and Ng, 2010; Miwa *et al.*, 2012, 2013; Vlachos and Craven, 2012; Ferraro *et al.*, 2013), but do not yield large performance improvements (Miwa *et al.*, 2012). An alternative strategy is to focus the use of algorithms on the sections of text more likely to be like the text in which the algorithms have been trained. For example, by running algorithms trained on Medline only on the sections of patents that discuss biological experiments.

The continuum between structured and unstructured data

Text mining has sometimes been defined as the mining of unstructured data, namely free text, as opposed to data mining, which is the mining of structured data (Feldman and Dagan, 1995). However, from the description of content sources made here, one can see that there is no abrupt separation between structured and unstructured content. Some content, such as normalized annotations, affiliation information and entries in clinical trial databases, can be fairly structured. Other content, such as structured abstracts, clinical records and clinical trial abstracts, has some degree of structure above that of a typical scientific abstract. For example, clinical trial abstracts tend to have a certain degree of organization around aspects such as inclusion criteria and enrollment, which allows for easier mining (Kiritchenko *et al.*, 2010). On the other hand, social media sources such as Twitter have a fluid grammar and a rapidly evolving language. Clinical records include many free-text annotations that are not grammatical (e.g., "blood pressure high"), non-standard abbreviations, and typos (Jensen *et al.*, 2012).

Sections add a layer of structure to a document. The full text of scientific articles is divided into sections such as introduction, methods, results and discussion (IMRaD). However, some content may be located outside of the section that is most appropriate (Agarwal and Yu, 2009). US patents have some typical sections such as abstract, claims, examples, and figures. Unfortunately, sectional information can be lost during the processing of a document and section boundaries may need to be detected. These boundaries can be recognized to some extent using rule-based methods for recognizing section headers, even though the formatting that helps human readers recognize them (font weight and size, paragraph spacing) might have been lost. The commercial tool Quosa, for example, is able to detect sections in scientific articles in some cases. The PDF-to-XML tool PDFX can recognize article sections in PDF documents (Constantin *et al.*, 2013).

The reference section is a highly structured part of a document. This section brings many redundant results in a text-mining exercise if the title of a highly cited reference contains relevant results. References can appear, in fact, as if they were an inherent part of a document's content and not just a link to another document. A way to avoid this issue is to eliminate altogether the reference section from documents. However, this has its drawbacks. The reference section can be mined for facts that are only found in the titles of references that are not part of the corpus. Thus, keeping the references increases the coverage of a corpus.

Even at the level of the vocabularies used, differences in level of structure are noticeable. Some vocabularies tend to be quite structured, such as the standard nomenclatures for chemical names, microRNAs, locus tags and mutations, while others are far less so.

The big picture is that text mining and data mining overlap. For example, patents contain tables (structured) which are described within figure descriptions and within the main text (unstructured). This is a more nuanced view than "the biomedical literature can be seen as a large integrated, but unstructured data repository" (Winnenburg *et al.*, 2008). Some prefer to even talk about Text & Data Mining (TDM), which would be a more general discipline that deals with content with different levels of structure. Publishers discuss text mining rights in the context of TDM licenses and the journal *Bioinformatics* publishes text mining articles under the section "Data and text mining."

References

Agarwal, S. and Yu, H. Automatically classifying sentences in full-text biomedical articles into Introduction, Methods, Results and Discussion. *Bioinformatics*. 2009;25(23):3174–3180.

Björk, B. C., Welling, P., Laakso, M., *et al.* Open access to the scientific journal literature: situation 2009. *PLoS ONE*. 2010;5(6): e11273.

Breiner, D. A. and Rodriguez-Esteban, R. Web Scraping Technology as a Cost-Effective Solution for News Alerting. Special Libraries Association, Pharmaceutical and Health Technology, Spring Meeting, Philadelphia, April 2013.

Canese, K. PubMed Celebrates its 10th Anniversary! *NLM Technical Bulletin*. 2006;(352):e5.

Caporaso, J. G., Deshpande, N., Fink, J. L., *et al.* Intrinsic evaluation of text mining tools may not predict performance on realistic tasks. *Pacific Symposium on Biocomputing*. 2008;640–651.

Clark, A., Körner, C. and Nielsen, H. P. From punched cards to apps and iPads. Fifty-five years of the P-D-R. *Business Information Review*. 2013;30(2):96–101.

Clegg, A. B. and Shepherd, A. J. Benchmarking natural-language parsers for biological applications using dependency graphs. *BMC Bioinformatics*. 2007;8:24.

Cohen, K. B., Johnson, H. L., Verspoor, K., Roeder, C. and Hunter, L. E. The structural and content aspects of abstracts versus bodies of full text journal articles are different. *BMC Bioinformatics*. 2010;11:492. doi: 10.1186/1471-2105-11-492.

Constantin, A., Pettifer, S. and Voronkov, A. PDFX: Fully-automated PDF-to-XML conversion of scientic literature. *Proceedings of the 2013 ACM symposium on Document Engineering (DocEng 2013)*. 2013;177–180.

Dahlmeier, D. and Ng, H. T. Domain adaptation for semantic role labeling in the biomedical domain. *Bioinformatics*. 2010;26(8):1098–1104.

Divoli, A. *Biomedical Text Mining Approaches: Applications in Protein Family Annotation* (dissertation). Manchester: University of Manchester, 2006.

Eder, J., Sedrani, R. and Wiesmann, C. The discovery of first-in-class drugs: Origins and evolution. *Nature Reviews Drug Discovery*. 2014;13(8):577–587.

Eriksson, R., Jensen, P. B., Frankild, S., Jensen, L. J. and Brunak, S. Dictionary construction and identification of possible adverse drug events in Danish clinical narrative text. *Journal of the American Medical Information Association*. 2013;20(5):947–953.

Feldman, R. and Dagan, I. *Knowledge Discovery in Textual Databases (KDT). First International Conference on Knowledge Discovery (KDD-95)*. Montreal, Canada, 1995.

Ferraro, J. P., Daumé, H. 3rd, Duvall, S. L., *et al.* Improving performance of natural language processing part-of-speech tagging on clinical narratives through domain adaptation. *Journal of the American Medical Information Association*. 2013;20(5):931–939.

Friedman, C., Kra, P. and Rzhetsky, A. Two biomedical sublanguages: A description based on the theories of Zellig Harris. *Journal of Biomedical Information*. 2002;35(4):222–235.

Golder, S. and Loke, Y. K. The contribution of different information sources for adverse effects data. *International Journal of Technological Assessment in Health Care*. 2012;28(2):133–137.

Gomes, B., Hayes, W. and Podowski, R. M. Text mining. In *In Silico Technologies in Drug Target Identification and Validation*, ed. D. Leon and S. Markel (pp. 153–194). Boca Raton, FL: CRC Press, 2006.

Jensen, P. B., Jensen, L. J. and Brunak, S. Mining electronic health records: Towards better research applications and clinical care. *Nature Reviews Genetics*. 2012;13(6):395–405.

Jiang, Y., Lin, C., Meng, W., *et al.* Rule-based deduplication of article records from bibliographic databases. *Database (Oxford)*. 2014;2014:bat086.

Jones, C. W., Handler, L., Crowell, K. E., *et al.* Non-publication of large randomized clinical trials: Cross sectional analysis. *British Medical Journal*. 2013;347:f6104.

Kabiljo, R., Clegg, A. B. and Shepherd, A. J. A realistic assessment of methods for extracting gene/protein interactions from free text. *BMC Bioinformatics*. 2009;10:233.

Kiritchenko, S., de Bruijn, B., Carini, S., Martin, J. and Sim, I. ExaCT: Automatic extraction of clinical trial characteristics from journal publications. *BMC Medical Information and Decision Making*. 2010;10:56.

Kulkarni, A. V., Aziz, B., Shams, I. and Busse, J. W. Comparisons of citations in Web of Science, Scopus, and Google Scholar for articles published in general medical journals. *Journal of the American Medical Association*. 2009;302(10):1092–1096.

Leaman, R., Wojtulewicz, L., Sullivan, R., *et al.* Towards internet-age pharmacovigilance: Extracting adverse drug reactions from user posts to health-related social networks. *Proceedings of the 2010 Workshop on Biomedical Natural Language Processing*. 2010;117–125.

Loging, W., Rodriguez-Esteban, R., Hill, J., Freeman, T. and Miglietta, J. Cheminformatic/bioinformatic analysis of large corporate databases: application to drug repurposing. *Drug Discovery Today*. 2011;8(3–4):109–116.

Martin, E. P. G., Bremer, E. G., Guerin, M., DeSesa, C. and Jouve, O. Analysis of protein/protein interactions through biomedical literature: Text mining of abstracts vs. text mining of full text articles. In *Knowledge Exploration in Life Science Informatics* (pp. 96–108). Lecture Notes in Computer Science. New York, NY: Springer, 2004.

McIntosh, T. and Curran, J. R. Challenges for automatically extracting molecular interactions from full-text articles. *BMC Bioinformatics*. 2009;10:311.

Miwa, M., Thompson, P. and Ananiadou, S. Boosting automatic event extraction from the literature using domain adaptation and coreference resolution. *Bioinformatics*. 2012;28(13):1759–1765.

Miwa, M., Pyysalo, S., Ohta, T. and Ananiadou, S. Wide coverage biomedical event extraction using multiple partially overlapping corpora. *BMC Bioinformatics*. 2013;14:175.

Primo Peña, E., Vázquez Valero, M. and García Sicilia, J. Comparative study of journal selection criteria used by MEDLINE and EMBASE, and their application to Spanish biomedical journals. The 9th European Conference of Medical and Health Libraries. 2004.

Pyysalo, S., Salakoski, T., Aubin, S. and Nazarenko, A. Lexical adaptation of link grammar to the biomedical sublanguage: a comparative evaluation of three approaches. *BMC Bioinformatics*. 2006;7(Suppl 3):S2.

Ramakrishnan, C., Patnia, A., Hovy, E. and Burns, G. A. Layout-aware text extraction from full-text PDF of scientific articles. *Source Code in Biology and Medicine*. 2012;7(1):7.

Rodriguez-Esteban, R. *Methods in Biomedical Text Mining* (dissertation). New York, NY: Columbia University, 2008.

Rodriguez-Esteban, R. and Iossifov, I. Figure mining for biomedical research. *Bioinformatics*. 2009;25(16):2082–2084.

Schmoch, U. Indicators and the relations between science and technology. *Scientometrics*. 1997;38(1):103–116.

Schuemie, M. J., Weeber, M., Schijvenaars, B. J., *et al.* Distribution of information in biomedical abstracts and full-text publications. *Bioinformatics*. 2004;20(16):2597–2604.

Searls, D. B. Mining the bibliome. *Pharmacogenomics J*. 2001;1(2):88–89.

Shultz, M. Comparing test searches in PubMed and Google Scholar. *Journal of the Medical Library Association*. 2007;95(4):442–445.

Swinney, D. C. and Anthony, J. How were new medicines discovered? *Nature Reviews Drug Discovery*. 2011;10(7):507–519.

Tanabe, L., Scherf, U., Smith, L. H., *et al.* MedMiner: An Internet text-mining tool for biomedical information, with application to gene expression profiling. *Biotechniques*. 1999;27(6):1210–1214, 1216–1217.

The Europe PMC Consortium. Europe PMC: A full-text literature database for the life sciences and platform for innovation. *Nucleic Acids Research*. 2014;pii: gku1061.

Tikk, D., Thomas, P., Palaga, P., Hakenberg, J. and Leser, U. A comprehensive benchmark of kernel methods to extract protein–protein interactions from literature. *PLoS Computational Biology*. 2010;6:e1000837.

Tomanek, K., Wermter, J. and Hahn, U. Sentence and Token Splitting Based on Conditional Random Fields. *Proceedings of the 10th Conference of the Pacific Association for Computational Linguistics* (PACLING 2007). Melbourne, Australia. 2007.

Van Landeghem, S., Hakala, K., Rönnqvist, S., *et al.* Exploring biomolecular literature with EVEX: Connecting genes through events, homology, and indirect associations. *Advances in Bioinformatics*. 2012;2012:582765.

Van Noorden, R. Trouble at the text mine. *Nature*. 2012;483(7388):134–135.

Van Noorden, R. Text-mining spat heats up. *Nature*. 2013;495(7441):295.

Van Noorden, R. Elsevier opens its papers to text-mining. *Nature*. 2014;506(7486):17.

Verspoor, K., Cohen, K. B., Lanfranchi, A., *et al.* A corpus of full-text journal articles is a robust evaluation tool for revealing differences in performance of biomedical natural language processing tools. *BMC Bioinformatics*. 2012;13:207.

Vlachos, A. and Craven, M. Biomedical event extraction from abstracts and full papers using search-based structured prediction. *BMC Bioinformatics*. 2012;13(Suppl 11):S5.

Weiss, G. M. Mining with rarity: a unifying framework. *SIGKDD Explorations Newsletter*. 2004;6:7–19.

Winnenburg, R., Wächter, T., Plake, C., Doms, A. and Schroeder, M. Facts from text: Can text mining help to scale-up high-quality manual curation of gene products with ontologies? *Briefings in Bioinformatics*. 2008;9(6):466–478.

Xu, S., Yoon, H. J. and Tourassi, G. A user-oriented web crawler for selectively acquiring online content in e-health research. *Bioinformatics*. 2014;30(1):104–114.

4 Integrating translational biomarkers into drug development

Jonathan Phillips

Drug development is notoriously slow and arduous in comparison to other high-tech industries. The raw nature of biology makes it very difficult to rapidly prototype and iterate in ways that are normally much faster in other technology spaces. Layer on top of the lengthy screening processes and multi-year, multi-million dollar clinical trials an enormous regulatory burden and you end up with development times in excess of 10 years on average from concept to product (Lipsky and Sharp, 2001). Contrast this to the smartphone industry, where a new hardware prototype can be in stores within nine months from the first mockup. Even faster, OS releases can sometimes turn around in 90 days. Granted, the improvement increments may be measured on different scales in pharmaceuticals and consumer electronics, but the key driver of innovation is the iterative cycle.

It is not surprising that the pharmaceutical industry has turned to computational approaches in order to compress the lag experienced between bench and bedside. With so much riding on a single molecule to perform in clinical trials and deliver the most promising product possible, early characterization of drug candidates can make or break the next blockbuster. Drug candidate identification and optimization has seen benefit from process automation, but prediction of which targets and drugs will perform well enough to provide a positive risk–benefit is still not possible.

Biomarkers are tools applied to study exploratory pharmaceuticals, which help scientists and clinicians better understand the trajectory of a developing drug. Historically, drugs were developed almost exclusively using empirical data to evaluate the risk–benefit profile. The use of biomarkers in drug development has driven a trend toward more quantitative, evidence-based drug development. This trend has also fueled the promises of personalized medicine and the companion diagnostics industry. Importantly, incorporating biomarker information throughout the drug development process has led to an increase in confidence to accelerate or abandon certain experimental therapeutics. The FDA's Critical Path Initiative has driven regulatory interest in using biomarker evidence to add greater confidence in bringing new therapeutics to unmet medical needs (Woodcock and Woosley, 2008). While the emerging trend to use biomarkers has provided plenty of buzz around precision medicine, sifting through all the information to find the right conclusions

Bioinformatics and Computational Biology in Drug Discovery and Development, ed. W.T. Loging. Published by Cambridge University Press.

can be a new challenge stemming from the mountains of data being generated in response. Here is where "Big Data" meets "Big Pharma."

In this chapter, the relationship between drug and biomarker concepts will be discussed. Common tools, both computational and wet-bench, will also be introduced. These tools are often used to co-develop drugs and biomarkers to be used to support more accelerated, evidence-based drug development. Occasionally, the accompanying biomarkers are converted into companion diagnostics or patient stratification tests. An overview of computational approaches for characterizing therapeutic targets and the drug properties needed for successful intervention (i.e., druggability) will also be shared. Once the drug therapeutic concept is well-defined, an accompanying biomarker concept is often developed. Specific categories of biomarkers and the types of questions they intend to answer are important aspects of developing and applying a biomarker concept to drug development. Three facets that determine the strength of a biomarker concept are biological relevance, technical feasibility, and translation potential. Computational approaches for maximizing each of these facets will be discussed here.

Combining drug development with biomarker information is only as useful as the way data are applied and interpreted. This chapter will conclude by discussing ways to fortify preclinical drug development projects with rich information to drive better decisions and improve confident delivery of the best candidates into the clinic.

Relationships between drugs and their biomarkers

Shared characteristics

Whether we are talking about molecular targets for drug concepts, or biomolecular markers to monitor various aspects of drug activity, we're talking almost exclusively about the same two categories of biological molecules: nucleic acids (e.g., DNA, RNA) and proteins (e.g., enzymes, receptors).

For drug targets, current dogma focuses on biological molecules, found naturally in the body, which have relevance to disease processes. Only druggable targets can, hypothetically, be influenced by pharmacological intervention.[1] Intervention is currently limited by the available modalities for delivering drugs successfully to the desired target. These options are expanding, but historically they were limited to synthetic chemicals, referred to as "small molecules" in the industry vernacular. These synthetic chemicals influence, often inhibiting but sometimes activating, the molecular role of the drug target to augment the disease or its pathogenic process. Within the past few decades therapeutic antibodies and recombinant replacement proteins, also known as "large molecules" or "biologics," have expanded the realm of which targets are druggable.

[1] The concept of "druggability" is discussed in more detail later in this chapter, with particular focus on ways to use common computational tools to better understand whether a target may be druggable.

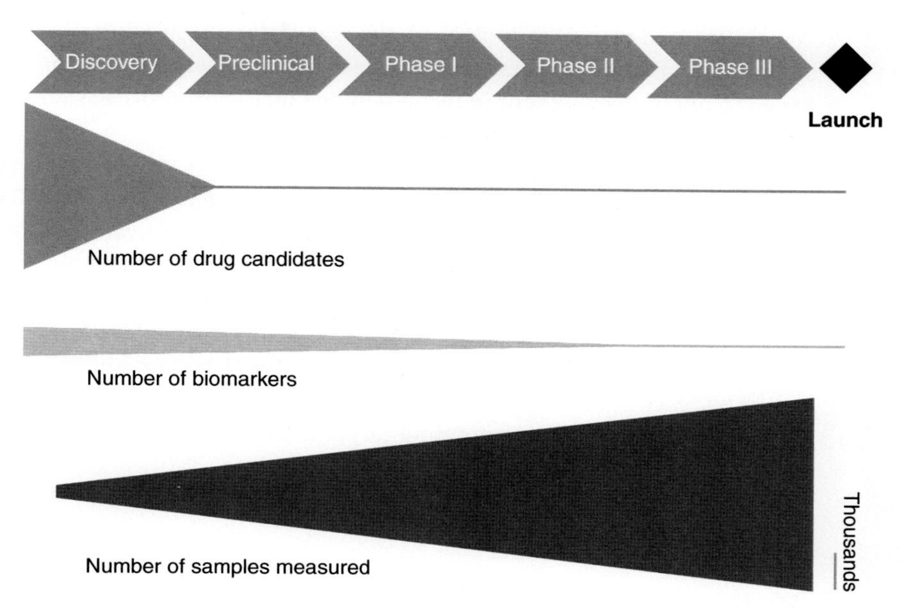

Figure 4.1 Parallelism between drug candidate screening and biomarker identification. The top row summarizes a typical drug discovery workflow. The second row represents the relative number of drug candidates screened where the height of the shape represents the numbers of drug candidates as they funnel down to one clinical candidate. The third row represents the relative number of exploratory biomarkers tested. The bottom row represents the relative number of biomarker samples tested.

For anyone hunting for new drugs, drug targets, or biomarkers, there is the ever-present balancing of biological relevance with technical feasibility. There is a dynamic equilibrium between these two competing features of molecular medicine. This happens at every scale of biomedical research. We use animal models because it is not feasible to do all facets of research on humans. As a result, we accept the limited biological relevance of animal models as they relate to humans. Not all aspects of this can be controlled, but if scientists do a good job of characterizing their tools and models, progress can be made. Similarly, at the microscopic level, we cannot shrink down to the scale of individual molecular or cellular processes. We have to accept the technical limits that come with making biological observations. The feasibility of measuring a transcription factor docking to its genomic DNA target is less than measuring the aggregate number of mRNA copies potentiated by that same docking event. Even for counting copies of mRNAs, we need the assistance of fluorescent probes and amplification steps to bring that information within the detectable range of our human senses.

To be clear, this chapter focuses on computational approaches for better understanding the underlying biology of drug targets and associated biomarkers. The collective design, development, engineering, screening, testing, and validation of thousands of drug candidates is a science in itself. Strategies and tools used to develop suitable drugs are topics for other entire books to discuss. Here, we will remain focused on extant biological raw materials to be used *in situ*. We will carry forward the assumption that state-of-the-art drug development approaches can be tuned to optimally intervene in the underlying biological processes we aim to deeply understand using the tools presented in this chapter.

One of the first steps to gain deeper understanding of the biological processes at the core of disease is to visualize the molecular components in action. Constructing a mechanistic model of known molecular interactions is invaluable for managing all of the possible intervention and monitoring opportunities within a disease process. Much of what can be assembled into a mechanistic model may be limited based on the state of the science and whether all the features of the process have been characterized. Even if major portions of the processes are known, the extent and confidence in resolution of their characterization may also be a limiting factor. A bedrock skill set for coordinating target and biomarker network biology is the ability to use and apply pathway analysis tools.

Common tools and concepts

Pathways

Biology is complex. Humans have an estimated average of 3.72×10^{13} cells (Bianconi *et al.*, 2013). The diploid human genome has 6×10^9 base pairs. Each base pair combination delivers 2 bits of data, meaning that each four-base pair combination in the genome represents a byte of information. This translates into approximately 1.5 gigabytes of storage per genome. Now we layer on top the countless diversity of terminally differentiated and continuously renewing cell types among those 3.72×10^{13} cells. The specific function each cell has within its tissue and organ drives the types of biochemical reactions occurring within and around that cell to support the overall homeostasis of the organism. Each of them are accessing up to 1.5 gigabytes of instructions to control, build or destroy all the components within the 1-nl volume of each of those cells, and the local space in between cells. Estimates of protein synthesis rates vary widely depending on a myriad of factors controlling expression, but consensus rates are in the ballpark of 30,000 methionine incorporations per cell, per second, where all proteins require at least one methionine to initiate translation (Li *et al.*, 2014).

If the genome is the cell's hard drive, the polymerases and ribosomes are the processors executing discrete tasks in the form of synthesizing proteins to manage the handling of variables that influence biological function. This translates into each cubic centimeter of tissue, containing at least 10,000 cells, as a 3000 MHz cluster with 150 terabytes of storage. The human body would then represent the combined processing power of a 117 petahertz (117×10^{15} Hz) cluster with 5.58 zettabytes (5.58×10^{21}) of total storage. Realistically, we are not going to be able to simulate that with current technology. To construct a sophisticated molecular model of the pathway interactions between each variant of those executable functions for each protein, RNA, and metabolite would be quite challenging.

While a computational model simulating all of the biomolecular activity in an intact mammal may not yet be feasible with current technology, we can leverage improvements in computational power to help chip away at this complexity problem. To drive toward a deeper understanding of these complex interactions building a mechanistic model is a standard starting point. In this section, we will discuss some of the computational tools that will allow better interface with the various types of components and interactions that make molecular pathways more comprehensible and manageable. We will keep the focus of this section on tools to build and visualize pathways. While pathway visualization and pathway analysis are somewhat inseparable, we will cover computational tools useful for pathway analysis later in this chapter.

Visualizing mechanistic models: tools

To better comprehend the inner workings of large dynamic systems, visualization tools are often useful. Edward Tufte has built a career on the "picture worth a thousand words" principle. His 1983 landmark book *The Visual Display of Quantitative Information* has helped many people deliver better messages using large amounts of data. Since then, Tufte has published several other books and essays on the topic of information visualization.

The pathway map is a fundamental tool for understanding molecular cellular biological processes. As new hypotheses about inter- or intracellular interactions occur, or as new data strengthen or weaken these hypotheses, the outcomes can be cataloged in a pathway map. Building and curating these sketched-out interactions help to fully understand the cellular role and context of the molecule within the system.

A variety of tools are currently available for constructing, editing and visualizing biomolecular pathways (see Table 4.1). Many of these tools behave similarly and essentially serve the same function. New tools are routinely being developed, making the diversity of tools available an advantage for those who desire specific, customized features to fit their purpose. The majority of pathway visualization tools are available as free, open-source solutions. We will explore two pathway visualization tools, highlighting some of the useful features, both unique to that particular platform and common to other pathway visualization platforms. One key consideration when selecting a pathway visualization solution is whether it has the import/export capability you need for the data sources you tend to handle. Nobody wants to start with a blank sheet and manually construct pathways or networked pathways. A goal for gaining familiarity with one or more pathway visualization solutions would be to develop facile workflows to readily handle the data sets most commonly encountered.

PathVisio

PathVisio is an open-source, Java-based pathway visualization and analysis platform. Features that PathVisio shares with other visualization platforms include the straightforward ability to draw, edit, and curate molecular pathways in a graphical

Table 4.1 Currently available and popular tools for biomolecular pathway visualization.

Name	Organizations	Website	Platform(s)	Good for …	PubMed hits[1]
BioTapestry	Cal Tech ISB, Seattle	biotapestry.org	Java	•Alternative display of networks •Longitudinal changes of networks	5 (52)[2]
Cytoscape	Cytoscape Consortium (Many)	cytoscape.org	Java	•Network visualization •Multiplatform database integration	423
GenMAPP	Many	genmapp.org	Windows	•Network visualization •Network analysis	90
MEGA	TMU, Tokyo Arizona State	megasoftware.net	Windows MacOS	•Comparative genomics •Phylogenetic trees •Pipeline analyses	60
Nbrowse	NYU	gnetbrowse.org	Java	•Network visualization	
NetPath	Johns Hopkins Institute of Bioinformatics (India)	netpath.org	Java	•Immune and cancer signaling pathways •Batch downloads	21
Pathview	UNC Charlotte	pathview.r-forge.r-project.org	R	•Integration with R •Visualization with analysis	3
PathVisio	Many	pathvisio.org	Java	•Network visualization •Network construction	13
Reactome	NYU EMBL-EBI Ontario Institute for Cancer Research Cold Spring Harbor Laboratory	reactome.org	Java Cytoscape	•Molecular interactions •Curated pathways	148
WikiPathways	Many	wikipathways.org	Java	•Quick reference •Comparative pathways	24

[1] PubMed hits are mentions of this particular tool via simple keyword search as of 27 Feb 2015.

[2] There are 52 self-cited publications listed on the BioTapestry website.

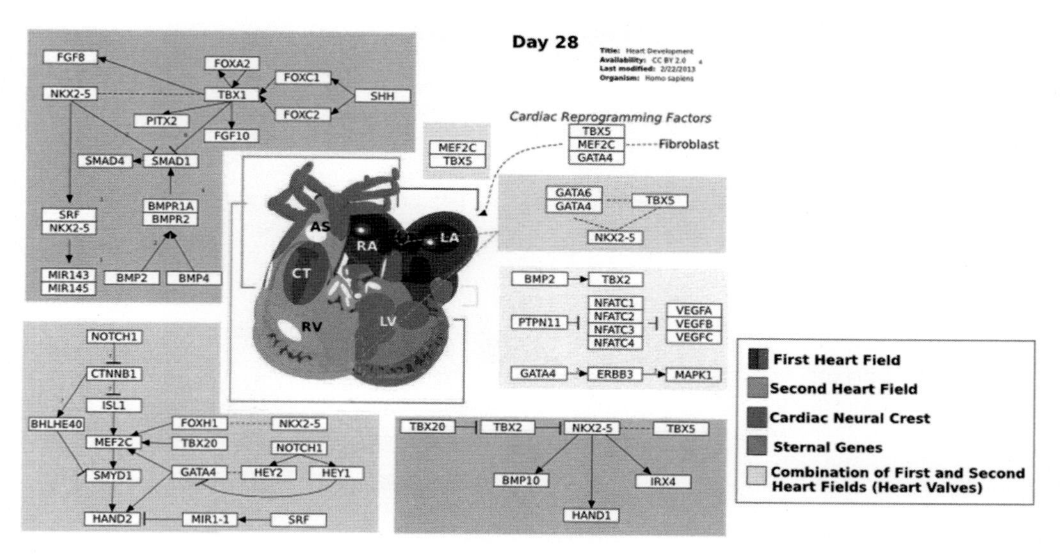

Figure 4.2 PathVisio. A screenshot from the PathoVisio platform describing an example of the pathway networks involved in cardiac physiology.

user interface (GUI) manner (Figure 4.2). Some of the flexibility to use PathVisio comes from the available suite of plugins compatible with the main platform. Many of these plugins allow access to previously constructed pathway data sets from other repositories. For example, there is a specific import/export plugin that allows data from PathVisio to be shared using the Biological Pathway Exchange (BioPAX) system, which is a standardized language intended to help simplify the exchange of pathway data (Demir *et al.*, 2010). Other plugins add analysis functionality to PathVisio. For example, open-source contributors have developed a plugin that allows for gene set enrichment to be conducted within PathVisio. This plugin is designed to identify over-represented pathways from loaded gene data sets. This particular plugin also incorporates import functionality for directly loading gene data sets from The Broad Institute's MSigDB. Finally, the flexible nature of PathVisio allows for the novel development of Java plugins to new or customized functionality. PathVisio curators will test user-submitted plugins for performance before being added to the plugin repository. Overall, PathVisio is a fairly simple tool, which makes it a solid option for beginners, with or without Java coding abilities.

Cytoscape

Cytoscape is another Java-based, open API pathway visualization solution, similar to PathVisio. Like PathVisio, Cytoscape allows users to load, manipulate, and visualize molecular interaction data. Cytoscape is a slightly more sophisticated solution, which has more options and flexibility to combine and manipulate multiple data sets (Figure 4.3). The Cytoscape ecosystem is more diverse; it includes a larger number of plugins (what are now called "Apps") and has a fairly vibrant user/developer community. Cytoscape seems to be the standard based on relative use, but it

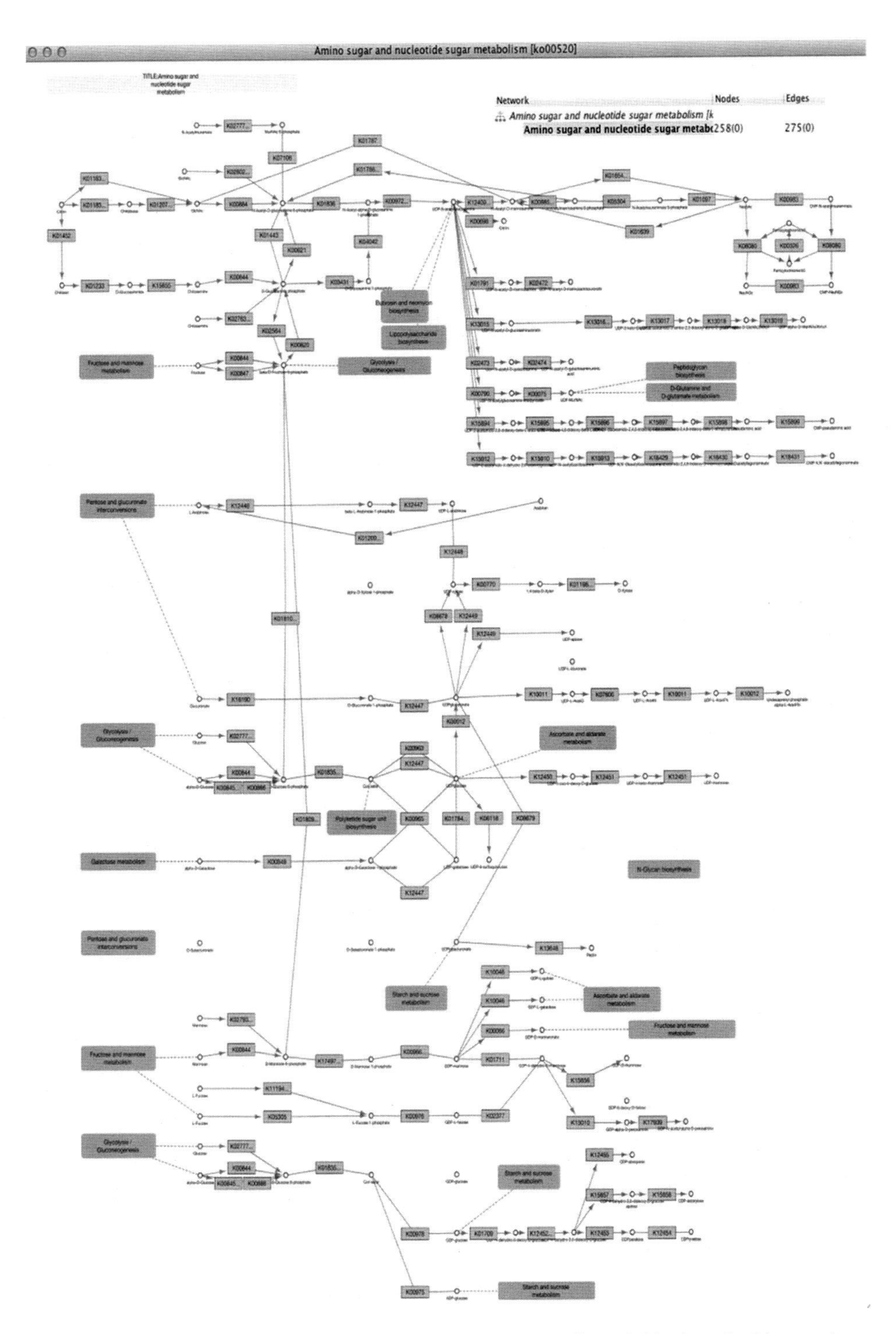

Figure 4.3 Example of the amino sugar and nucleotide sugar metabolism. A black and white version of this figure will appear in some formats. For the color version, please refer to the color plate section.

may be a bit cumbersome to new users who are trying to start navigating and managing pathways in a visual manner.

Drug properties and optimization

Which is the holy grail of drug development: the ability to identify new drug targets, or the ability to construct a drug that will best antagonize/agonize a drug target? Identifying new drug targets is only as useful as the ability to engineer the drug that will most safely modify disease. These two concepts go hand-in-hand, so we are really talking about the simultaneous capacity to characterize pharmacologically relevant targets *and* the drug properties that will give greatest control over the pathophysiology.

Druggability

There are countless estimates of what constitutes the definitive set of druggable targets. The answer to this problem is constantly in flux because the factors for determining druggability are constantly changing, improving, and under debate. The main consideration for druggability remains therapeutic modality, which most often boils down to whether you can drug the target with a small molecule or biologic. With new experimental modalities ranging from RNA silencing and gene therapy to nanoparticles and cellular engineering, it will not be surprising if the term "druggable genome" ceases to exist (Figure 4.4).

Modalities: general considerations

Small molecules

A tremendous amount of science has been invested in understanding what constitutes the best "drug-like" properties of small molecules. Physicochemical properties such as solubility and permeability are cornerstones for narrowing down chemical space that might make it as a possible drug. A landmark manuscript in 1997, republished in 2001, nicely summarized the main chemical features useful for identifying novel drugs (Lipinski *et al.*, 2001). This paper establishes the small molecule "rule of five," which spotlights hydrogen bond acceptors, hydrogen bond donors, molecular weight, calculated Log P, and solubility as the most influential parameters for drug-like molecules. A variety of computational approaches are available for characterizing synthetic organic chemical structures for their drug-like properties. Many of these approaches are laid out nicely in the Lipinski paper with a wealth of references to other strategies for identifying drugs with a high probability of success.

In this section, we will focus on the general characteristics of biological targets that may make them more amendable to drugging with a small molecule. Among all marketed drugs, the targets of these drugs can be classified into a handful of biomolecular categories. The top five drug target classes include: G protein-coupled

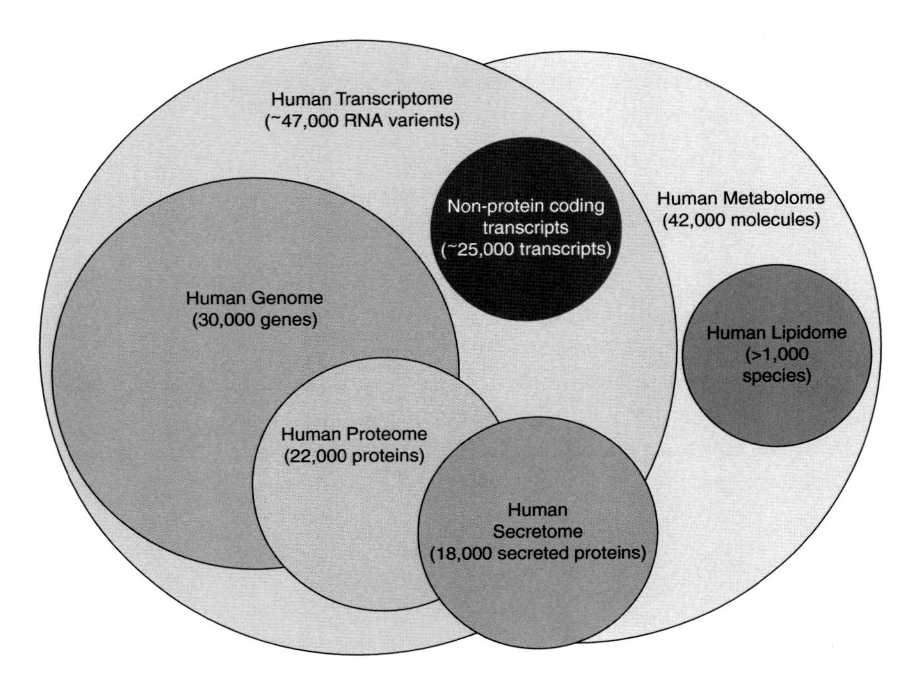

Figure 4.4 The many varieties and flavors of 'omics. Defining boundaries for what consists of an 'ome has been a moving target for decades. The discovery of new biomolecular species has added to the complexity as new bins of molecules have been described. As we become more sophisticated in the way we access and manipulate the various biomolecules within living systems, the way we define their biochemical taxonomy also changes. Perhaps the most variable 'ome has been the druggable genome. This is intuitive as we consider the many advances in the way drugging modalities have evolved.

receptors (GPCRs), kinases, ion channels, proteases, and phosphatases (Hopkins and Groom, 2002). Clearly, enzymes represent the vast majority of drug targets. Activity focused into a catalytic pocket for most enzymes makes them prime drug targets for small-molecule inhibitors. Most drugs behave as inhibitors by outcompeting the natural substrate or ligand of the target. In many cases the drug mimics the natural substrate, but with greater binding strength and lower, if any, biological activity.

Because most druggable targets belong to a particular gene family that shares similar structural and functional characteristics, the notion that other members of that family may be druggable is reasonable. However, not all druggable targets are related to disease function. Many drug targets are not dominantly associated with disease and may only play a minor role in pathogenesis. An estimated 10% of all druggable targets have sufficient influence over a disease process to be candidates for disease-modifying intervention (Walke *et al.*, 2001). Many of the estimates of disease-modifying genes are based on knockout mouse data. These models are a good start; however, the true pathogenesis may not be fully reflected by a single gene elimination. We know that much of biology, including dysfunctional biology, is a coordinated effort among biomolecular pathway components. Some of these components synergize to confer phenotype, so understanding which of the many contributors to pathophysiology may require understanding the mutual contributions of many pathway components. How to simultaneously drug the key components to gain the maximal effect becomes exponentially complex. Co-inhibition of multiple targets by a single agent can be a challenging undertaking, but leveraging conserved gene family structure has benefited several drug concepts. For example, the triple kinase inhibitor nintedanib (Vargatef® or OVEF®) is a single agent capable of potently inhibiting VEGF receptors, FGF receptors, and PDGF receptors. Nintedanib works by competing with ATP in the binding pocket of these receptors. The conserved structure of the ATP binding pockets allows for multiple targets, but the precise structure of nintedanib also allows for selective inhibition of only these families of growth factor receptors.

Large molecules: immunogenicity predictions, accessibility, proximal targets with improved druggability

Large molecules predominantly include engineered monoclonal antibodies. Therapeutic antibodies operate under a similar intent as small molecules where they are often designed as inhibitors to compete with a ligand or substrate binding site. In addition to classical inhibitory functions, large molecules can be designed to perform a variety of other functions.

While some small-molecule approaches take advantage of the conserved nature of common catalytic structures within gene families, large molecules can be engineered to bind multiple targets with very diverse structures. Monoclonal antibodies are considered bivalent because they naturally contain two binding sites within their CDRs (complementarity-determining regions). The CDRs are the parts of the variable chains in antibodies that confer binding specificity and give antibodies their exquisite selectivity. Antibodies can be modified to switch out the variable regions such that they contain binding sites for different targets. Furthermore, exotic antibody-like formats are constantly being developed that bring additional binding sites into the structure affording valences greater than the natural two found in standard monoclonal antibodies. As a result, the engineering of therapeutic antibodies can now include multiple targets with precisely selected binding affinities to give optimal drug properties for a given therapeutic concept.

These and other novel protein engineering approaches allow drug developers to expand the boundaries of the druggable target library.

Biomarker concepts in drug development

Biomarker classifications

In this section, we will review the common categories of biomarkers and the general questions they intend to answer. The goal at the end of this section is to be able to

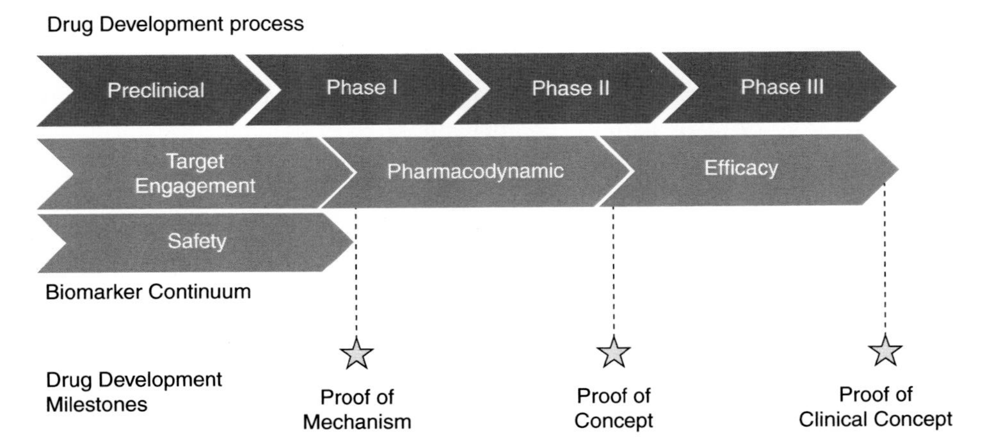

Figure 4.5 A proposed biomarker continuum identifying the key biomarker classes necessary to build a drug development program. Starting with target engagement and safety at the translational interface between preclinical and clinical research, these biomarkers are necessary to understand the drug–target pathway relationship. Pharmacodynamics links the molecular relationship between the drug target and its influence on the overall pathophysiology. Pharmacodynamic biomarkers can help further validate target pathways and reinforce confidence in a disease process. Efficacy biomarkers are simply surrogates for clinical assessment of the overall disease state. Efficacy markers are perhaps the most challenging because they rely on empirical categorization and characterization of a disease diagnosis.

understand the key features of high-quality biomarkers, how they are used to assess drug performance, and ways the data may be used to provide deeper context for enriching outcomes of preclinical studies (Figure 4.5).

Target engagement

A good target engagement biomarker will provide the most proximal readout of whether a drug candidate is physically interacting with its intended target. A suitable target engagement biomarker will often be a direct measurement of the target itself. For example, the marketed drug Avastin® (bevacizumab) is a therapeutic monoclonal antibody that selectively neutralizes the vascular endothelial growth factor-A (VEGF-A). The way this drug works is by binding as much available VEGF-A as possible so that it cannot interact with its natural receptor VEGFR2. To understand whether bevacizumab is interacting with its target, VEGF-A, an assay to detect blood levels of VEGF-A can be used as a target engagement biomarker. If the total amount of VEGF-A is tracked over time in a dosed subject, a rapid increase in total VEGF-A would be observed followed by a gradual decrease. These data make sense because VEGF-A is taking on the longer half-life of bevacizumab, therefore increasing the total amount of VEGF-A in the system. In this way, the observation of increased total VEGF-A in a bevacizumab-dosed subject provides positive evidence of target engagement.

These measurements can be further characterized to better understand the interactions between bevacizumab and VEGF-A by splitting out the total amounts into respective fractions of free and bound drug. Examining target engagement in this manner allows scientists to understand how dose relates to coverage of the target. In many cases, drug developers want to augment as much of the disease pathway as possible, ostensibly by covering the maximum amount of target. In other cases, there is a rationale for low or moderate target coverage. Balancing safety and efficacy by limiting unintended effects stemming from exaggerated pharmacology is a common reason for limiting the coverage window. Assessing target engagement during *in vivo* experimentation is also a way to provide translational context to drug–target interactions. Orthologous targets expressed in animal models may interact with drugs differently than their human counterparts. In some cases where the animal target is homologous to the human target, the biological environment may be different, which may also influence the drug–target relationship. These interactions may not be purely biophysical in nature, but may also be affected by stoichiometric or allosteric features within the organism.

In the bevacizumab example described above, measuring VEGF-A gives an indication of the status of the drug–target interaction. One common temptation is to use a downstream marker as an indicator of whether the experimental drug is hitting its target. To extend the bevacizumab example, using phosphorylated VEGFR2 as a target engagement biomarker would be inappropriate. While a change in phosphorylated VEGFR2 is a desired outcome of bevacizumab treatment, this measurement does not provide direct information about the interaction between bevacizumab and VEGF-A. Measuring phosphorylation status of VEGFR2 may reveal information about the interaction between VEGFR2 and its ligands, which would be an indicator of pathway signaling. In the next section we talk about monitoring pathway engagement, or what may also be referred to as pharmacodynamics.

Pharmacodynamics

Pharmacodynamic biomarkers provide a way to monitor steps that reside between altering the function of the drug target and altering the course of disease. This is a broad classification of biomarker that can actually be broken down further depending on the resolution of pathway engagement desired (or feasible). The boundaries defining a pharmacodynamic biomarker start with target engagement and range through target modulation, pathway response, cellular response, physiological response, pathology modifying, ending at efficacy. Depending on the objective of the drug's therapeutic concept, any of these intermediate steps can be considered pharmacodynamic. Scientists will often cloud the definition by interchangeably referring to any of these observations as evidence of target engagement or pharmacodynamics.

These semantics are less important when the objective is narrowly defined, the biology is well understood and the quality of measurements is highly reliable. More often, scientists are exploring uncharted territory and will adopt a panel approach for interrogating one or more of these types of pharmacodynamic markers in hopes of achieving one or two of these criteria. Much of this boils down to the technical feasibility of making a measurement. Also important is the ability to connect the pharmacokinetic and pharmacodynamic (PK/PD) relationship as a way to predictably understand pharmacology.

Using the earlier example of bevacizumab engaging its target VEGF-A, we can start to understand the pharmacodynamics of bevacizumab by measuring and studying downstream pathway influences. Because bevacizumab aims to augment the VEGF-A pathway, we can use the natural receptor to VEGF-A as a pharmacodynamic indicator of pathway engagement. In fact, many anti-angiogenenic therapeutic approaches targeting VEGF biology use VEGF receptors as pharmacodynamic biomarkers. Phosphorylation of VEGFR2 by VEGF-A is a well-described process (Ferrara *et al.*, 2003). Assessing the phosphorylation status of VEGFR2 would be considered a direct indicator of pharmacodynamics. With these types of measurements, considered with pharmacokinetic data and target engagement data, a fuller picture can be developed with respect to the way dose level, dose schedule, and systemic exposure to bevacizumab can alter the VEGF-A pathway.

Efficacy

Efficacy biomarkers are perhaps the classification of biomarker most people are familiar with and most are comfortable understanding. Efficacy biomarkers also tend to be the most subjective and variable because they frequently rely on clinical metrics, experiential judgment, or a combination of lines of evidence. The holy grail of biomarkers is to identify the ideal molecular "surrogate" that is able to substitute for these more subjective classifiers. The inherent nature of an objective measurement substituting for a subjective parameter lends to considerable variability and error. Much of this stems from our limited understanding of most pathological processes and subsequent ability to develop the correct diagnostic categories.

Not all efficacy biomarkers are soft measures. The ultimate goal of any drug is to resolve disease and the assessment of this goal is only limited to the ability to assess disease status. For example, many solid tumors are directly measurable in size by conventional imaging techniques. Tumor volume and a more temporal measurement of tumor growth can be quantitatively assessed by imaging for most solid tumors. The limits of these approaches are bound by the limits enabling accurate measurement (e.g., resolution of MRI). In this example it is currently not possible to assure that all cancerous cells have been cleared of the body. It is also not clear what defines a concerning number of cancer cells. Is one cancer cell too many? Are 10,000 cancerous cells too many? Questions like these make medicine an imperfect art to practice and not a precise science.

Drug safety and biomarker translation

Biomarkers for drug safety are a distinct category because they may perform the job of monitoring on-target, off-target or most often a combination of on- and off-target adverse effects. Many drug safety biomarkers are applied routinely in toxicology and clinical safety studies as clinical chemistry panels. These panels include biomarkers that are not designed to work based on a particular drug mechanism, but monitor common target organs susceptible to drug-induced injury.

Presenting biomarker translation together with drug safety is a natural discussion because much of drug safety is predicated on non-clinical safety studies. The importance of animal models in drug safety is critical. Enormous weight is given to toxicology studies and patient safety is non-negotiable in developing new experimental therapies. The need for drug safety biomarkers to translate into clinical use is paramount. As such, much attention has been given to the predictive nature of novel drug safety biomarkers.

The Predictive Safety Testing Consortium (PSTC) is a pre-competitive collaboration between health authorities (FDA, EMA, PMDA) and some of the largest pharmaceutical companies. The PSTC represents the first public–private effort to bring formal regulatory review of evidentiary standards needed to globally accept new biomarkers as translational drug development tools. Some of the first products of this collaboration include the regulatory qualification of several drug safety biomarkers (Dieterle *et al.*, 2010; Sistare *et al.*, 2010). These early regulatory qualifications were founded in non-clinical models first and designed to encourage exploratory use in clinical studies to build more experience and help better understand their true translational nature. It is common knowledge that non-clinical models can have widely varying limitations on the translatability of outcomes into patient outcomes. This carries down to the molecular cellular level, where drug injury processes can also vary, making some biomarkers work well in animals, while others only work well in humans. This is the goal of these accelerated public–private efforts to review emerging data on the latest approaches for drug safety monitoring.

Patient delimiting

Previously, when we discussed efficacy biomarkers, we also discussed some of the difficulties of binning patients into a disease category or measuring disease severity. Some patient delimiting biomarkers face similar challenges depending on the question the biomarker is designed to answer. There are many synonyms for patient delimiting biomarkers. Some of the more common terms used include "patient stratification," or worse yet "predictive biomarkers." If you are reading this book, then you likely have an appreciation for computational technology and terminology. Within that realm, the term "delimiter" refers to one or more characters, often in sequence, used to identify boundaries or separate areas of information. Bringing this term into biomarker development brings a parsing feature that helps characterize patient populations in a way that guides whether they may be more or less responsive to a particular therapeutic approach. Biomarkers are, in essence, delimiters, which help us better understand the behaviors of disease processes and what characteristics certain individuals may possess that would more precisely categorize their potential responsiveness to therapy.

Disease diagnostics are essentially biological definitions of disease. Clinicians apply these boundaries to either rule out a diagnosis or confirm one. Diagnoses are often confused as being definitions for a particular disease, when the opposite is more accurate: the disease is often defined by the method of diagnosis. For example, simple home pregnancy tests measure the amount of human chorionic gonadotropin (hGC) in the urine. If a woman tests positive for hGC then they are "diagnosed" as pregnant;[2] women who test negative are classified as not-pregnant. The absolute truth of the hGC status is used to determine whether the test subject is pregnant or not. This is not to be confused with the diagnostic power of the assay that measures hGC, which is subject to analytical error.[3] Analytical errors can result in false positives or false negatives, but the assumption is that the true status of a diagnostic biomarker will determine the medical diagnosis.

Symptoms are a special class of patient delimiter. These features may be used as inclusion or exclusion criteria to better control a clinical trial. Very often, multiple symptoms are used to aid in classification of a complex disease state because of their low specificity. Diseases do not really care how humans classify them, just like living organisms do not care about their taxonomic position in phylogenetic trees. Classifying diseases is purely a human invention. The diagnostics that are associated with disease classification are nearly as much a human invention as they are a function of nature (Figure 4.6).

Competitive advantages for doing these approaches

A robust biomarker program associated with an innovative drug therapeutic approach is paramount. The main competitive advantage of having quality mechanistic biomarkers throughout drug development is to speed up the development process by increasing confidence in the drug's performance at each stage. In this way, drug developers get a snapshot of the direct influence that drug pharmacology has on augmenting disease and associated disease mechanisms. For much of the history of drug development, novel drugs have relied on empirical observations to determine the probability that a new drug will satisfy a risk–benefit balance. Translational, precision medicine now relies on molecular diagnostics, high-resolution imaging, and other technologies used as biomarkers to quantitatively measure the direct influences of the disease process.

Deepening biological relevance

As new medical information is being discovered every day, scientists can take advantage of that context to develop tools for exploring new therapies. Here we discuss how to access deeper relevance of the molecular components of disease. By identifying these components and applying a check of technical feasibility to determine whether they are accessibly measurable, we can discover new biomarkers.

[2] While the premise of this chapter is to identify disease-augmenting drug targets and their biomarkers, this author does not believe pregnancy is a diagnosable disease. To be clear, the diagnostic in this example does perform as a delimiter to properly classify women who are pregnant and women who are not pregnant.

[3] Diagnostic potential and methods for evaluating them will be discussed later in this chapter under Mathematical Interpretation.

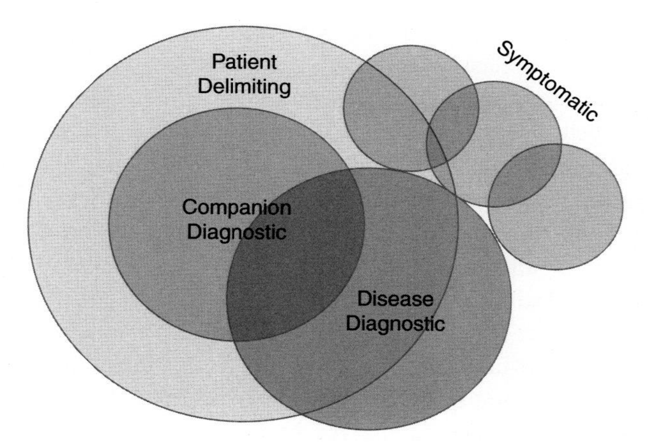

Figure 4.6 Venn diagram of patient delimiting, diagnostic, companion diagnostic, and symptomatic indicators. One of the greatest medical challenges is precise and adequate description of the pathophysiological processes that define a disease. With any luck, a disease can be defined by a single, dichotomous factor that can be determined as positive or negative. This is where much optimism for pharmacogenomics comes from; where single nucleotide polymorphisms (SNPs) may determine a disease status. Precision medicine relies on such indicators, but the vast majority tend to be symptomatic and only able to partially describe some aspects of an overall disease process.

Those components that also have a functional impact on disease and are druggable may also be drug targets.

Text analytics

This book contains separate sections dedicated to text analytics, which you are strongly encouraged to explore (Chapter 3 and Appendix I). In this segment, we focus on strategies for applying text analytics and text mining for identifying molecular components of disease and associated components to be considered biomarkers. Text mining is a powerful way to validate or generate hypotheses based on current scientific thinking. One example of a commercially available text-mining platform is the Linguamatics I2E package. This solution uses what is termed "natural language processing" (NLP) to extract relevant chunks of information from written documents, such as scientific manuscripts or abstracts. One advantage NLP brings to literature searching is the ability to geographically associate key terms to other concepts. For example, the distance between key terms gives a strong relational context and allows for previously unrealized relationships between concepts to be discovered (Gan *et al.*, 2013).

One of the first, fundamental tools for successful text analytics is an ontology structured to index your literature corpora with common terms useful for identifying biomarkers. One of the most comprehensive medical ontologies is the SNOMED-CT ontology (www.ihtsdo.org/snomed-ct). Any ontology can be customized to improve the balance between indexing performance and specificity. Splitting the applied ontology into focus areas helps construct contextual queries. Here, five separate ontological themes are emphasized in this chapter for the purpose of focusing text analytics queries onto biomarker discovery and disease relationships: phenotypes, body structures, targets, drugs, and species. "Phenotypes" is a collection of diseases and diagnoses that can be used to help describe a particular medical condition. For example, many pathology terms should be contained in this ontology. This ontology will help expand or contract searches around particular diseases or pathological processes. "Body structures" allows queries to be focused on a particular anatomical feature. It also allows the flexibility of searching for anatomical structures that may be affected by a particular drug, gene, or pathophysiology. This is useful in drug safety for identifying possible off-target effects. "Targets" is a compendium of all known genes, which forms the general basis for target and biomarker biology. While this is not a comprehensive approach for biomarker discovery, it offers an accessible means for rapidly testing hypotheses with nucleic acid analysis, either at the mRNA or DNA level. Finally, "species" is a complete list of common laboratory animal species names and synonyms to cross-reference translational queries. This ontology helps the understanding of whether certain biology is conserved in a particular model species.

A sixth ontology, which should be part of most text analytics approaches, is a relational ontology (relationships). "Relationships" is an index of words that helps give directionality to a query to determine how the two are expected to interact (see Figure 4.7). The ontologies and their structures can be customized to suit specific needs. These guidelines are only intended to aid in focusing the most relevant features of large ontologies in order to gain deep context in the areas of drug target and biomarker discovery.

Various types of text-analytics queries can be performed to mine out information to provide deeper context. One example is a query constructed to identify where a candidate drug target may be expressed in the body. This would cross-reference the specific drug target against a "body structures" ontology. To expand on this example, two illustrations can be considered: one where an expected result is confirmed by text analysis, and another where a non-obvious result is discovered.

In these two illustrations, we are interested in confirming and further understanding the overall tissue relevance of two molecular targets that are intended to be used in immunology indications. These targets are expected to be on immune cells and function to suppress inflammatory responses. In the first illustration we examine the protein B-cell activating factor (BAFF). We know BAFF has a role in immunological function and we are interested in inhibiting its function to improve inflammatory diseases. To confirm the role of BAFF in immune function and see if there are any potential off-target tissues or cell types, we query the available tissue or cell types which are being discussed in the context of BAFF in the current biomedical literature. A corpus of all abstracts within the US National Library of Medicine's PubMed database was indexed against a partial SNOMED ontology, structured similarly as described above. A query was then constructed to identify relationships between BAFF and categorize hits according to the "body structures" portion of the ontology. In this example, the relational terms included known laboratory methods that could be used to make these observations. Methods such

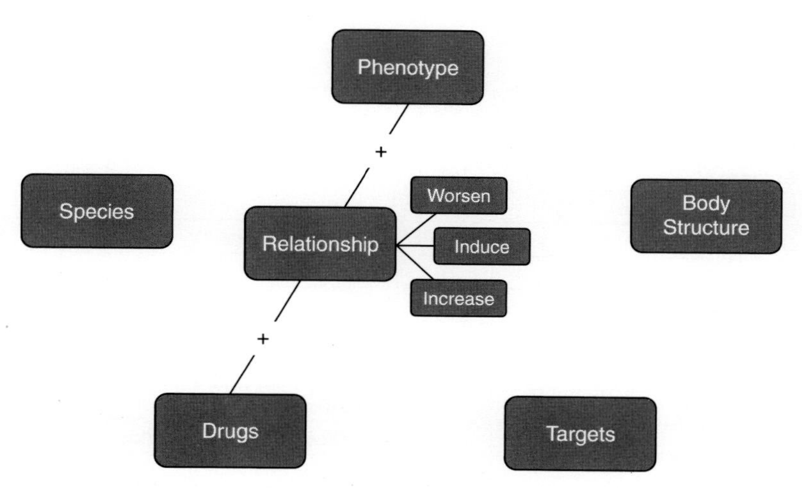

Figure 4.7 Categories of ontologies useful for building text analytics infrastructure related to mining out information relevant to biomarker discovery and translational medicine.

as immunohistochemistry and *in situ* hybridization were specifically used to extract direct connections between the target and associated tissue or cell types contained within the "body structures" ontological tree (Figure 4.8). Looking at the tabulated results, we can see that immune-related tissues and cell types are indeed most frequently associated with BAFF in the literature. Further validation can be found at the top of the list, where B lymphocytes are most frequently associated, approximately four times as often as the next most mentioned structure, the spleen.

In the second illustration, we examined fractalkine receptor (GPR13) in the same way. Similar results are expected compared to BAFF; GPR13 is a chemokine receptor with a predominant role in T cells and monocytes in response to fractalkine signaling, resulting in aggressive chemotaxis toward the signal. We see from the results that macrophages are ranked as the fourth most frequent unique hit, which supports the pharmacological concept. What are more prevalent are descriptions of central nervous tissues and cells. The brain, neurons, and spinal cord are all listed in the top five most frequent hits. Following the citations delivered from this analysis, we can learn that GPR13 has a role in the development of the nervous system as evidenced by knockout mouse phenotypes. We also see that GPR13 has an influence on microglial function. These results may suggest a potential unintended effect in the nervous system if GPR13 was targeted ubiquitously. A well-curated list of references to GPR13 knockout models can be found at www.informatics.jax.org/marker/MGI:1333815.

We can also use this approach for interrogating biomarker information. Queries can be assembled that help identify biochemicals associated with disease or pharmacological activity. Likewise, these queries can be structured to identify biomarker characteristics, such as endogenous, normal, and disease-associated levels. Revisiting the angiogenesis theme, we can use VEGF-A as an example again here. In this case, a group of assay developers needs to understand the anticipated levels of VEGF-A in plasma or serum. A query was constructed to mine out text references

B-cell activating factor (BAFF)

Tissue	uHits
B-Lymphocytes	**641**
Spleen	**156**
Lymphoid Tissue	**118**
Bone Marrow	**116**
Salivary Glands	79
Lymphocytes	**58**
Appendix	**54**
Macrophages	**54**
Monocytes	**54**
Lung	49
Blood	**44**
Dendritic Cells	**44**
Germinal Center	**44**
T-Lymphocytes	**32**
Serum	**20**
Palatine Tonsil	**18**
Plasma Cells	**18**
Bone Marrow Cells	**16**
Leukocytes	**16**
Endothelium	14

Fractalkine Receptor (GPR13)

Tissue	uHits
Brain	**323**
Neurons	**259**
Astrocytes	**243**
Macrophages	**204**
Spinal Cord	**195**
Neuroepithelial Cells	**142**
Microglia	**117**
Endothelial Cells	**122**
Leukocytes	**103**
CD4-Positive T-Lymphocytes	**102**
Kidney	101
Dendritic Cells	**98**
Lung	91
Spleen	**90**
Neuroglia	**85**
Neural Stem Cells	**71**
Endothelium	58
Liver	57
Breast	52
Heart	51

Figure 4.8 Target localization within tissues by semantic relationships using natural language processing. Specific proteins and their synonyms were queried for relationships connected with laboratory methodologies used to directly observe the presence or absence of target expression in tissues and cells. Unique hits (uHits) were then curated, summed and ranked by frequency of uHits to derive a consensus tissue of interest. Bolded results are expected biological sites of expression for a particular drug target. Bolded gray results are potential "off-target" locations indicating a tissue or cell type that is not within the scope of the intended pharmacological mechanism of a targeting drug concept. A black and white version of this figure will appear in some formats. For the color version, please refer to the color plate section.

to VEGF-A and its relationship to the terms "serum" or "plasma." A subontology of quantitative terms was also used to improve precision. Terms such as "concentration," "amount," "volume," and various synonyms such as "ng/ml" or "mg/dl" were used to focus the results on analytical data. The goal was to determine a consensus measurement of VEGF-A concentration in blood samples. These quick analyses yielded several dozen citations indicating the endogenous concentration of VEGF-A in serum is approximately 30 pg/ml. This query can be adjusted to broaden the same type of search by selecting a branch of the ontology containing VEGF-A one level higher to capture other members of angiogenic growth factors. Re-running the query with this minor adjustment resulted in identifying handfuls of citations with the putative endogenous concentrations of VEGF-C, VEGF-D, ANGPT1, ANGPT2, and soluble VEGFR2 in blood matrices.

These examples and illustrations are intended to guide thinking around computational methods for text analytics and text mining to derive deeper context when searching for relevant information during biomarker discovery and development efforts. These approaches can also make translational or reverse translational adaptation of methods and biology much simpler by including query features that characterize the "species" involved in the extracted text mentions.

Pathway analysis (next-nearest neighbor)

Pathway analysis becomes useful again when searching for drug-relevant biomarkers. When a drug target has been identified, the search for biomarkers begins with understanding the pharmacological concept behind drugging that particular target. In many cases, neutralizing or inhibiting a target is expected to have several downstream signal transduction consequences. Learning what those specific consequences are can be readily assessed using pathway analysis and looking at the up- and downstream neighbors (i.e., next-nearest neighbor) in the putative pathway. In this section we discuss a few representative tools to quickly interrogate next-nearest neighbors by performing simple pathway analysis to identify molecules expected to interact with a given target.

Biomarker concept qualification

Just like drug targets, biomarkers must be qualified across two main dimensions: biological relevance, and technical feasibility. Up until now, this chapter has focused on how to derive the most biological relevance for a biomarker concept as it relates to a given drug targeting mechanism. It is right and appropriate to prioritize the most biologically relevant biomarkers before evaluating technical feasibility of measuring those markers. The temptation to approach biomarker development in the reverse order is tremendously limiting, stifles innovation, and leaves many opportunities for competitive advantage on the drawing board.

Analytical validation

The term "validation" is often over-used. In the regulated drug development environment, at least one health authority, the FDA, has clearly separated the intended use of the term validation to mean analytical validation of methods used to measure endpoints along the drug development continuum. The FDA's *Guidance for Industry: Analytical Procedures and Methods Validation for Drugs and Biologics* is a good starting place when advancing a technical feasibility evaluation for biomarker protocols (www.fda.gov/downloads/drugs/guidancecomplianceregulatoryinformation/guidances/ucm386366.pdf). As with many other areas in drug development, validation is a progressive event that can be useful short of fully accomplishing a complete validation. The term "fit-for-purpose" has been used to identify minimal evaluation conditions to declare a method useful, but not robustly characterized to the extent that may be regulatory ready. Fit-for-purpose methods are those that can deliver informative results to be trusted for internal decision-making. These results are often considered by regulatory authorities as supportive evidence when considering a decision, but rarely accepted as standalone evidence for pivotal endpoints. The aim of qualifying and validating analytical methods progressively is to speed up decision-making while continuously evaluating and improving the feasibility of those procedures. As new information or technology becomes available during the qualification process, progressive, fit-for-purpose validation allows the flexibility to adapt and improve to more appropriate procedures.

Ultimately, the level of final validation should be dependent on the end use of the biomarker. If the biomarker is accepted as exploratory and only contributes incremental evidence for internal decision-making, then a minimal quality control evaluation would be necessary to the extent that the procedures are deemed acceptable to deliver the answer needed. As endpoints become more critical for demonstrating proof of a significant milestone in a drug program's progress, the level of quality validation increases and discussions with health authorities around the proper controls and tolerances will factor into the true technical feasibility of a particular protocol.

Mathematical interpretation

Once biomarkers are identified and initial results start to look promising, then the questions start coming. How do we know we really have a change? At what level can we declare something is actually happening? What makes us think this biomarker is any better than what is already used in the clinic? To answer questions like these, we turn to some mathematical analyses to help make objective comparisons. In this section we will discuss contemporary and emerging approaches for evaluating biomarker performance. The two general areas of discussion will include:

(1) statistical analysis to evaluate classification performance of a biomarker test; and
(2) mechanistic modeling of experimental measurements to simulate study conditions and predict the influence of interactions and pharmacological intervention.

Design of experiments

Design of experiments (DoE) is an area of statistics focused on effective experimentation. Formal DoE is becoming an important aspect of drug development as the conduct of studies and cost of time becomes more expensive. Applying DoE aims to reduce time and cost of experimentation, limit biasing risks, and increase the scope of results. When seeking new biomarkers, applying a trial and error approach is incorrect. Too frequently, researchers are attracted to the idea of "fishing expeditions" in search of a biomarker that will "work." In this approach, a variety of random biomarkers, sometimes disguised by being referred to as a "panel of biomarkers," are tested, and based on the results some may be kept, or more experiments, with more subjects and more time points, are conducted until the desired outcome is observed.

An incremental improvement over trial and error would be to apply traditional experimentation where one variable at a time is modified while all others are fixed. While traditional experimentation aims to find optimal results, there is no guarantee that the right combination of factors will be discovered using this method. Also, it is nearly impossible to detect interactions between factors. Interactions between factors frequently mask the correct results when varying one factor at a time. Varying factors individually also increases the number of iterations needed, often to an infeasible level, further limiting the ability of finding the correct solution.

To further improve on access of the full experimental space while optimizing design parameters, a three-phase, iterative approach with statistical evaluation at each phase is recommended. Most research and development processes take this iterative approach, but not often enough are all the available data used to inform next steps. The first phase is screening, which aims to maximize the number of factors (biomarkers) while limiting the number of levels to be tested. In the screening phase a maximum level and a minimum level should be identified and worked within. The screening approach is used to identify influential biomarkers while avoiding the loss of any promising biomarkers.

The second phase is modeling, where the influential biomarkers are identified and the metrics that determine influence can be made. The decision on whether the measurements can be made is a question of technical feasibility and constitutes an important decision at the end of the screening phase, before the modeling phase can formally begin. Statistical expression of responses and interactions of variables can be represented mathematically (i.e., by equations, mathematical models), once the candidate biomarkers have been identified by screening and deemed technically feasible to measure. The number of biomarkers is further reduced based on simulated information derived from the models and optimization or confirmatory experiments can be designed based on the modeled results. Information learned from subsequent experiments can be used to further optimize processes and experimental designs.

This approach of screening (piloting), modeling and optimizing can be a continuous, iterative process. Information learned at each step can be incorporated to powerfully inform the design of the next experiments, which will then further tune the system to get to the correct answer more quickly and more accurately.

Power analysis

Applying the principles of DoE is preferred, but not always reality. At a minimum, to avoid unnecessary work, a power analysis of endpoints to be measured on a pre-designed study should be conducted. A frequent occurrence in translational biomarker application to drug development is the proliferation of “exploratory biomarkers” to studies intended for use in support of regulatory filing. Phase I enabling good laboratory practice (GLP) toxicology studies, pivotal clinical trials, and studies requested by regulators should only contain measurements that can be proven supportive for delivering decision-level evidence. Exploratory studies should be kept separate, where learning about unknown assay performance and biological influence is done in a hypothesis-driven manner.

Power analysis helps avoid irrational proliferation of measurements, even if the measurements are deemed established from a mechanistic or translational standpoint. Power analyses help pre-determine decision thresholds by linking the sample size from a given study to a required magnitude change in a particular biomarker based on the biological and technical variability of that biomarker. The more desired use of power analysis is to provide confidence in support of a measurement and any conclusions derived from those measurements. For these reasons, a power analysis of each biomarker proposed for each study should be conducted as part of practicing the principles of DoE.

Predictive models: applying biomarker data to decisions

Biomarkers are essential components of prediction models where multiple lines of evidence can be incorporated into a decision-guiding algorithm. Risk prediction models are one type of prediction model where the risk of an adverse event is determined by an algorithm that considers a variety of information, including biomarker measurements. Together with qualitative evaluation, risk prediction models are often constructed to deliver objective reasoning to guide healthcare decisions. In drug development, risk prediction models can be useful for delivering mechanistic information related to the drug mode of action as it relates to the probability of adverse events occurring. Information such as plasma drug level, clinical chemistries and physical observations can be used to determine whether exposure (or over-exposure) to a drug may present a safety risk.

Multivariate prediction model construction is becoming a common practice in all areas of drug development. With the arrival of precision medicine, the introduction of new biomarkers and tools for accessing biomarker information demands innovative approaches for assimilating the massive amounts of data. There is a natural attraction to the promise of enlightening insight provided by novel biomarkers. This attraction also leads to the proliferation of biomarker measurements on drug development studies. Adding large amounts of disparate data actually works against the predictive power of multivariate models due to overfitting (Subramanian and Simon, 2013). Likewise, direct correlation analysis is not considered an appropriate method of determining biomarker performance (Figure 4.9). Here we discuss a handful of popular approaches for evaluating the diagnostic potential and diagnostic improvement a biomarker may have on a predictive model.

Ultimately, predictive models should aim to obey the parsimony principle, which simply suggests that models should only be as simple or as complex as they need to be. If a multivariate regression model can explain an outcome with only three factors, then using four factors violates parsimony. Likewise, if a calibration curve is expected to be linear, then it should be fitted to a linear function. Fitting a data set that is expected to be linear to a quadratic function would violate parsimony.

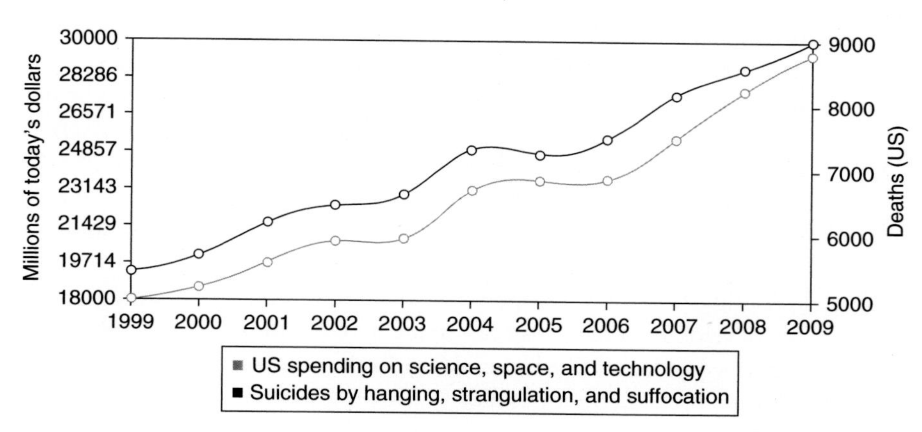

Figure 4.9 Warning against direct correlation analyses. Spurious correlations are everywhere. For example, Tyler Vigen makes an art out of identifying spurious correlations from random data sets. In this illustration, Tyler performs a correlation analysis between United States spending on science, space, and technology from 1999 to 2009 against the number of suicides by hanging, strangulation, or suffocation. Is US technology spending a biomarker to predict how many may die from self-strangulation, hanging, or suffocation? Artful interpretation is a plague among biological sciences and blunt correlations are too often recruited to give credibility to dubious conclusions. (Creative Commons free license.)

Regression models

Regression modeling aims to explain relationships among variables. There are numerous types of regression models that can be used to connect relationships between factors (biomarkers) and outcomes. When we specifically discuss regression modeling in the context of biomarkers for drug development we aim to determine how a biomarker measurement can coordinate information within the large space that exists in the relational understanding of how a drug can influence disease. We can imagine using the continuum of biomarkers in drug development as stepping stones. These stepping stones can be converted into objective milestones as sufficient experimental evidence is collected to demonstrate proof (Figure 4.5). Regression models can aid in tying these events together. For example, if a key objective of a therapeutic concept is to cover greater than 90% of a drug target with an inhibitory drug, then target engagement measurements automatically have a conditional goal of demonstrating that 90% of the target is engaged by drug. Proof of mechanism studies can now use regression modeling to connect relationships between drug dose, plasma exposure and target engagement to predict what dose level will result in the desired 90% engagement level. As we progress from target engagement to pharmacodynamic effects, we are then asking the next incremental question about the relationship between the drug occupying the target and its influence on the target pathway biology. A new regression model, or set of models, can then be developed with the aim of determining whether the drug concept is valid to engage the disease pathway. Quantitative relationships can be built connecting the drug dose level and influence on the downstream components of the disease pathway. These relationships provide feedback on the robustness of dose selection and the variability of population responses at the level of the molecular pathway. Finally, surrogate efficacy biomarkers can be incorporated to determine the ultimate influence on disease state. This step is likely the most challenging, because individual biomarkers are not often available as single diagnostic delimiters of healthy/disease classification. As drugs approach late-stage development, the information available becomes more qualitative and descriptive, and this represents new challenges for incorporating into predictive models.

Taking this stepwise approach of co-developing biomarker concepts alongside the development of a drug candidate makes it more manageable and enables faster, higher-confidence decision-making. In the following sections, we take a closer look at some tools gaining popularity for progressively evaluating the performance of biomarkers in their ability to classify outcomes.

Receiver operator characteristic (ROC)

The hallmarks of biomarker performance boil down to sensitivity and specificity. Sensitivity is simply an indicator of whether a biomarker can deliver true positives. This makes sense because if a biomarker is not sensitive it would not detect a change and predict an outcome. If a biomarker is too sensitive it would only deliver positive signals, suggesting the biomarker is not specific. As a result, we see how specificity is intimately related to sensitivity. Specificity is an indicator of a biomarker's ability to deliver true negative results. If a biomarker is over-sensitive, delivering all positive results, then it cannot discriminate between subjects that do not have a particular condition.

For example, we want to treat patients who have high blood pressure with an experimental drug aimed at reducing blood pressure. Preclinical studies have shown that at very high exposures to this experimental drug there is evidence of liver damage, or hepatotoxicity. Physicians who will be responsible for conducting the Phase I clinical trials would like to have a tool they know can sensitively and specifically detect cases of hepatotoxicity as they start to characterize this drug in humans. Fortunately, there are some well-accepted biomarkers of hepatotoxicity available. For the first-in-human study, the physicians recommend using blood serum levels of aminotransferases to identify cases of hepatotoxicity. Before we apply this as a safe use condition, we want to evaluate the performance of serum aminotransferases in the preclinical animal studies. To get a composite snapshot of sensitivity and specificity, a receiver operator characteristic (ROC) curve is very useful. In this example, the way ROC analysis would be applied would be to start by identifying the true cases of hepatotoxicity in the toxicology studies. The gold standard of microscopic histopathological liver changes is used to determine "truth" in this instance. Animals that do or do not have hepatotoxicity are dichotomized into positive and negative categories. Serum aminotransferase levels are then ranged from highest to lowest observed within the data set. The positive and negative cases are then compared against the range of aminotransferase levels and a probabilistic output is generated, indicating the likelihood a particular level would result in a positive classification.

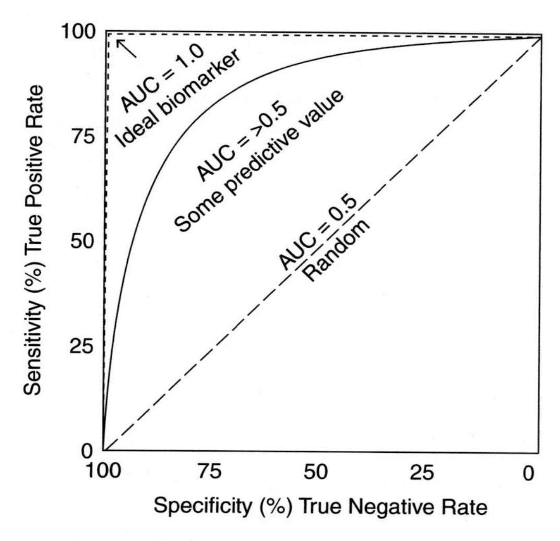

Figure 4.10 Receiver operator characteristic (ROC) analysis. ROC analysis provides a composite view of diagnostic potential when comparing a particular factor (biomarker) against cases observed (effect). Both sensitivity and specificity can be viewed at any threshold within the test range. Optimal threshold can also be identified at the point where the least number of total false positives and false negatives are observed. This threshold will also identify the maximum possible combined sensitivity and specificity. Area under the curve (AUC) is another metric used to quantitate performance. AUC values above 0.5 (greater than random) indicate positive predictivity, with 1.0 being ideal.

A ROC curve allows investigators to scan all possible levels of sensitivity and specificity. Comparison of the area under a ROC curve (AUC) also allows us to compare individual biomarker performance for sensitivity and specificity. The probability that a particular biomarker measurement is useful for identifying an outcome is another way to think about AUC. Typical AUC values range from random (0.5; a coin flip) to ideal (1.0), meaning that the biomarker can perfectly discriminate all positive and negative cases (Figure 4.10). An asymptotic analysis of a ROC curve can find the maximal sensitivity and specificity combination and deliver a threshold value for optimal biomarker performance based on that data set. This readout will also be related to the analytical performance of the assay used to make the biomarker measurement. If there is significant technical variability in the assay, this will be reflected in the evaluation of whether the biomarker can discriminate cases of hepatotoxicity, as in this example.

In many cases, a biomarker will either perform more sensitively or more specifically. Many investigators seek a single biomarker to deliver the most sensitive and most specific readout. More often, results are more reliable if a very sensitive biomarker is combined with a very specific biomarker to deliver that optimal combination. The challenge is determining the best combination of biomarkers and the best way to combine their use for a particular application. Here is where we need to adhere to the parsimony principle.

Net reclassification (NRI and IDI)

In 2008, Michael Pencina and colleagues proposed two new methods for assessing the value added to a predictive model when including a new parameter (i.e., biomarker) in the prediction (Pencina *et al.*, 2008). The challenge they aimed to meet was to develop statistical tools tailored to evaluate the usefulness of new biomarkers. Traditional statistical methods delivering a *p*-value seem too liberal, especially for studies with large sample sizes. To assess how a new biomarker contributes to correct classification, Pencina and colleagues developed two new approaches: net reclassification improvement (NRI) and integrated discrimination improvement (IDI).

NRI and IDI are introduced here following a review of ROC because the concepts are related. Comparison of ROC AUC can provide useful information about the relative performance between two biomarkers. For example, the PSTC successfully qualified seven novel kidney biomarkers for improved detection of drug-induced kidney injury based on a comparison of novel biomarker performance versus standard serum creatinine and blood urea nitrogen (BUN) levels (Dieterle *et al.*, 2010). These comparisons were done by direct evaluation of ROC AUC values. A downside of comparing ROC AUC values is the often low-magnitude difference between AUCs, which tends to minimize the meaning of differences detected by this method. Pencina and colleagues recognized the trend to use reclassification tables when comparing the inclusion or exclusion of a new biomarker for diagnostic improvement. To extend this trend, a formal accounting of movement due to reclassification from inclusion or exclusion of a biomarker was proposed as a way to track overall value added by biomarker information. The process of accounting for all biomarker-mediated reclassification is termed net reclassification improvement.

Let's revisit the previous hepatotoxicity example using aminotransferases as a biomarker of drug-induced liver injury. In this instance, we seek to compare the value of adding a cut-off of twofold increase in serum total bilirubin (TBIL) on top of a threefold upper limit normal (ULN) aminotransferases. This approach of combining aminotransferase information with TBIL is actually part of an accepted method, also known as Hy's Law, for evaluating cases of drug-induced liver injury. How do we know that adding TBIL to aminotransferase information brings more value to classifying cases of hepatotoxicity, as compared to using aminotransferase information alone? Creating a reclassification table allows us to see how many cases move from incorrect classification to correct and vice versa when TBIL data are included (Figure 4.11). To calculate NRI, the reclassification of subjects with and without hepatotoxicity is compared and the number of cases reclassified as improved or diminished is accounted for. In this example we have three risk categories. This approach can be used for any number of categories greater than one, but the overall concept is to track the direction in which certain cases are moving. Here we identify improved cases as those where a higher risk prediction is identified for those subjects with hepatotoxicity and lower for those subjects without hepatotoxicity. Similarly, a diminished result occurs when using TBIL delivers a lower risk

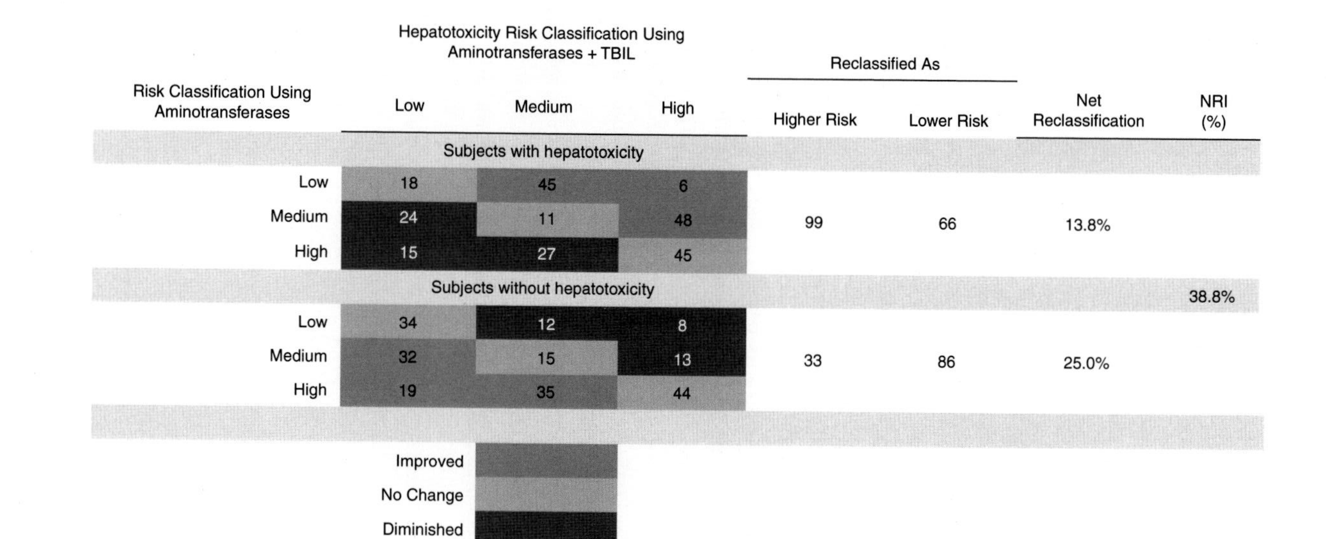

Risk Classification Using Aminotransferases	Hepatotoxicity Risk Classification Using Aminotransferases + TBIL: Low	Medium	High	Reclassified As: Higher Risk	Lower Risk	Net Reclassification	NRI (%)
Subjects with hepatotoxicity							
Low	18	45	6				
Medium	24	11	48	99	66	13.8%	
High	15	27	45				
Subjects without hepatotoxicity							38.8%
Low	34	12	8				
Medium	32	15	13	33	86	25.0%	
High	19	35	44				

Figure 4.11 A reclassification table is useful to visualize more detail about cases with improved or diminished classification when biomarker information is added. Net reclassification can be directly computed from these data. In this example, we look at improvement of classifying risk for hepatotoxicity using classical biochemical indicators, aminotransferases, and total bilirubin. This type of analysis helps determine if combinations of biomarker information add value and significance can be tested statistically.

category for those with hepatotoxicity (increased false negatives) and a higher risk category for those without (increased false positives). In this case, we see that the NRI identifies that adding TBIL improves overall diagnostic confidence by 38.8%.

NRI treats data in a dichotomized fashion, such as treating subjects as either with or without hepatotoxicity in the previous example. IDI differs by recalculating the risk for each subject using each risk prediction model with and without TBIL, which gives weight to the actual change in risk. An IDI analysis aims to find the overall impact TBIL has on the risk prediction model across all possible thresholds.

Conclusions

The approaches described in this chapter are intended to be illustrative starting points to be used as a foundation for building a robust biomarker system in support of innovative drug concepts. We discussed shared relationships between drug targets and associated biomarkers useful for developing those target concepts. Tools used to identify good biomarkers can also be useful in evaluating new drug targets. In fact, many drug targets are good biomarkers, and vice versa; the difference depends on the questions and approaches applied.

This chapter also proposes a generalized framework for integrating biomarkers into drug development programs using themes to step from one development milestone to the next. This approach allows for opportunities to observe drug performance in the direct context of its therapeutic hypothesis. Knowledge derived from these intermediate observations can serve to accelerate iterations, feed back into drug optimization, and enable quick kills while avoiding unnecessary, time-consuming empirical experimentation.

Finally, we bring together the experience derived from developing a quality biomarker program into powerful, evidence-driven decision applications. Many translational medicine organizations stop short of this important step due to the risks of not having enough information or confidence to allow an entire program to ride on the outcomes of novel measurements. These guiding principles, which are derived directly from solid application of the scientific method, allow separation of drug development organizations that will progressively move forward, apart from those that will grind slowly into the shadows of healthcare.

References

Bianconi, E., Piovesan, A., Facchin, F., *et al.* An estimation of the number of cells in the human body. *Annals of Human Biology*. 2013;40(6):463–471.

Demir, E., Cary, M. P., Paley, S., *et al.* The BioPAX community standard for pathway data sharing. *Nature Biotechnology*. 2010;28(9):935–942.

Dieterle, F., Sistare, F., Goodsaid, F., *et al.* Renal biomarker qualification submission: a dialog between the FDA-EMEA and Predictive Safety Testing Consortium. *Nature Biotechnology*. 2010;28(5):455–462.

Ferrara, N., Gerber, H. P. and LeCouter, J. The biology of VEGF and its receptors. *Nature Medicine.* 2003;9(6):669–676.

Gan, M., Dou, X. and Jiang, R. From ontology to semantic similarity: Calculation of ontology-based semantic similarity. *Scientific World Journal.* 2013;2013:793091.

Hopkins, A. L. and Groom, C. R. The druggable genome. *Nature Reviews Drug Discovery.* 2002;1(9):727–730.

Li, G. W., Burkhardt, D., Gross, C. and Weissman, J. S. Quantifying absolute protein synthesis rates reveals principles underlying allocation of cellular resources. *Cell.* 2014;157(3):624–635.

Lipinski, C. A., Lombardo, F., Dominy, B. W. and Feeney, P. J. Experimental and computational approaches to estimate solubility and permeability in drug discovery and development settings. *Advances in Drug Delivery Review.* 2001;46(1–3):3–26.

Lipsky, M. S. and Sharp, L. K. From idea to market: The drug approval process. *Journal of the American Board of Family Practitioners.* 2001;14(5):362–367.

Pencina, M. J., D'Agostino, R. B., Sr., D'Agostino, R. B., Jr. and Vasan, R. S. Evaluating the added predictive ability of a new marker: From area under the ROC curve to reclassification and beyond. *Statistics in Medicine.* 2008;27(2):157–172.

Sistare, F. D., Dieterle, F., Troth, S., *et al.* Towards consensus practices to qualify safety biomarkers for use in early drug development. *Nature Biotechnology.* 2010;28(5):446–454.

Subramanian, J. and Simon, R. Overfitting in prediction models – Is it a problem only in high dimensions? *Contemporary Clinical Trials.* 2013;36(2):636–641.

Walke, D. W., Han, C., Shaw, J., *et al. In vivo* drug target discovery: Identifying the best targets from the genome. *Current Opinions in Biotechnology.* 2001;12(6):626–631.

Woodcock, J. and Woosley, R. The FDA critical path initiative and its influence on new drug development. *Annual Review of Medicine.* 2008;59:1–12.

5 Computational phenotypic assessment of small molecules in drug discovery

William T. Loging and Thomas B. Freeman

The need for computational methods to characterize a new therapeutic compound's potential secondary pharmacology early in the drug development process is becoming increasingly important. The United States Food and Drug Administration (FDA) acceptance rate of new chemical entities (NCE) used in the treatment of human diseases has been unchanged over the past 60 years despite dramatically increasing investment in the past two decades (Munos, 2009). Multiple factors have contributed to this decrease in NCE approvals per unit investment including increased FDA standards, therapeutic approaches addressing more complex diseases, as well as issues with patient pharmacogenomic diversity.

As more small molecules are designed using combinatorial libraries and computational/structural biology methods, new and disparate forms of chemical matter are being produced for NCE consideration. These novel compounds do not have a history of associated side-effect profiles that established chemical matter have (e.g., penicillin analog). A retrospective analysis shows that nearly 30% of all new NCE failures in the year 2000 were attributed to problems with clinical toxicology, far more than any other single reason (DataMonitor, pharmaceutical report). From 2008 to 2010 there were 108 reported Phase II failures; of those reporting reasons for failure, 19% were reported due to clinical or preclinical safety issues (Arrowsmith, 2011a). At later stages in the development pipeline, combined successes in Phase III and submission have fallen to approximately 50%, with 83 failures between 2007 and 2010. Twenty-one percent of the failures across all therapeutic areas are due to safety issues (Arrowsmith, 2011b). Accordingly, while some progress has been made, it is critical that compound safety issues be addressed as early in the discovery pipeline as possible to reduce costly late-stage attrition.

The basic premise of drug toxicology is simple, but is complicated by the sheer size and complexity of the human proteome (Waring *et al.*, 2015). Compounds, or their metabolites, that interact with the desired target protein can also bind to and alter the activity of other "off-target" proteins. Many times, these proteins can have their activity altered without significantly affecting normal human physiology.

Bioinformatics and Computational Biology in Drug Discovery and Development, ed. W.T. Loging. Published by Cambridge University Press.

However, a protein's altered activity can lead to a change in a metabolic or signaling pathway critical to normal physiological function and hence to toxicological effects. Thus, identification of these "off-target" proteins and understanding the role they play in the human body is important. Toxicologists have identified many such proteins, an example of which is the HERG potassium channel (Thomas *et al.*, 2004; Leishman and Rankovic, 2014; Villoutreix and Taboureau, 2015): compounds that inhibit HERG can prolong the heart's QT cycle, producing an abnormal rapid heart rate. While there are several *in vitro* and *in vivo* toxicology assays available to monitor for compound interactions with HERG, this is but one example in a system of many thousands of interactions.

In recent years, several methods and databases have become available to identify toxicological effects using a variety of chemical structure, screening, and cheminformatic analyses. These can be broken down into three groups: those utilizing quantitative structure–activity relationship (QSAR) models, those utilizing complex statistical analyses such as neural networks, and those utilizing pharmacology databases. Methods that utilize QSAR models are predictive computational programs such as Deductive Estimation of Risk from Existing Knowledge (DEREK; Sanderson and Earnshaw, 1991; Patlewicz *et al.*, 2003), Toxicity Prediction by Komputer Assisted Technology (TOPKAT), and CASETOX (Greene, 2002; Senese *et al.*, 2004). These tools predict the potential mutagenicity and carcinogenicity of a query compound based on a model of known carcinogens. Other approaches include novel chemogenomics/chemoproteomic approaches (Niculescu *et al.*, 2004; Schuffenhauer and Jacoby, 2004) that link compound structure to potential toxicity incorporating additional models, and/or including QSAR. These models use complex statistical analyses of compound structure to predict toxicity and lethality. Pharmacology databases, such as BioPrint (Krejsa *et al.*, 2003), correlate experimentally determined associations between compounds' activity and known clinical secondary pharmacology using a broad range of target protein assays. We sought to overcome the limitations of previous programs by creating a novel approach that makes use of chemo-proteomics data.

Here, we report on a new computational biochemogenomic method, the computational phenotype assessment (CPA), that links compounds to a phenotypic/toxicological ontology via screening and literature databases and can be used to quickly triage compounds early in the development process. We benchmarked our method using 153 compounds with known secondary pharmacologies and found that it is able to correctly identify a significant number of events with a low rate of false positives.

The CPA method

The compound of interest is compared to all compounds found in a screening database using a Tanimoto similarity search with a threshold score of 0.6 to find structurally similar compounds. The screening database is again queried looking for proteins inhibited by the full set of similar compounds with an IC_{50} of less than 500 nM

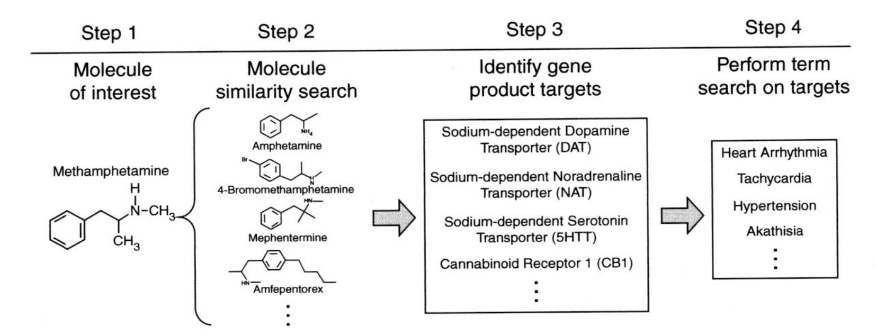

Figure 5.1 Computational phenotypic assessment (CPA) functions by bridging several different techniques, such as small-molecule structure searching, text analytics, and statistics.

(for this benchmarking analysis, the original compound is not used in the analysis). These targets are then analyzed against a phenotype ontology using a binomial statistical test in a Bernoulli experiment to determine the *p*-value significance of the protein vs. ontology term match (Figure 5.1). Finally, all ontological terms are displayed with their associated *p*-value. A retrospective analysis of the predicted phenotypic term for each of the test set small molecules indicated that a corrected benchmark *p*-value of significance, or a *p*-value cutoff of 0.01, should be used.

Test sets

To benchmark the process, we identified 153 drugs withdrawn from the market. The United Nation's document entitled "Consolidated List of Products Whose Consumption and/or Sale Have Been Banned, Withdrawn, Severely Restricted or not Approved by Governments" (United Nations, Department of Economic and Social Affairs, 2003) contains a list of substances that have been shown to cause unwanted secondary pharmacology with human administration as determined by clinical trials and population studies (not all drugs were withdrawn globally). Three hundred seventeen small molecules were taken from the list and their chemical structures analyzed in terms of the Lipinski "Rule of Five" (Lipinski *et al.*, 1996) to identify those with the most drug-like properties. Of these compounds, 153 were identified as being drug-like and labeled for internal use. This final list includes compounds ranging from sulfanilamides (antibiotic) to doxepins (CNS anti-depressant; see Supplemental Table 5.1) that was used to determine the success rate.

False positive test set

A second set of 120 currently available prescription drugs listed by the FDA as having few to no observable secondary events was used as the false positive test set (see Supplemental Table 5.2). The clinical trials, as well as on-label information, of these drugs were reviewed to identify compounds with few or no known major secondary pharmacology.

Table 5.1 Examples of the phenotypic/toxicological ontology terms. Each term is assigned to one of three separate therapeutic areas: cardiovascular (CV), gastrointestinal (GI), and musculoskeletal (MUSCULO).

Disease term	Therapeutic disease area
Hypotension	CV
Hypertension	CV
Angina	CV
Shock	CV
Gastritis	GI
Vomiting	GI
Emesis	GI
Constipation	GI
Akathisia	MUSCULO
Arthritis	MUSCULO
Dystonia	MUSCULO
Osteoarthritis	MUSCULO

Screening database

CPA uses a test-screening database containing >1000 assays of >750 protein interactions with more than 300,000 unique compounds. The IC_{50}, or concentration at which of the protein's activity is reduced 50% in the presence of compound, was included in the screening test set. We chose proteins that provided a broad representation of druggable gene families across the human proteome, ranging from of G-protein coupled receptors (GPCRs), kinases, and proteases, enzymes such as lipoxygenases and lyases, to DNA/RNA replication machinery, and signal transduction proteins. Of course, any major small-molecule database that contains binding metrics could be used in this process, which could increase or decrease the efficiencies of the CPA approach.

Phenotype/toxicology ontology

More than 13 million Medline abstracts were analyzed for relationships between proteins and toxicological terms. This approach has been described previously (Baker and Hemminger, 2010; Campbell *et al.*, 2010; Loging *et al.*, 2011). In our analysis, we used 95 ontological terms that were divided into three distinct therapeutic divisions: cardiovascular, gastrointestinal, and musculoskeletal (Table 5.1).

Tanimoto similarity searching

The Tanimoto search method calculates the similarity between two ligand fingerprints (Willett *et al.*, 1998). Each molecule, the reference molecule ("search

Table 5.2 The most highly predictable toxicological phenotypic terms in the CPA analysis.

A Cardiovascular

Indication	# Known	# Correct	% Correct
Tachycardia	46	43	93
Heart arrhythmia	67	58	87
Hypotension	42	35	83
Hypertension	48	39	81
Angina	29	22	76
Coronary heart disease	29	20	69
Myocardial infarction	19	13	68

B Gastrointestinal

Indication	# Known	# Correct	% Correct
Gastritis	9	9	100
Irritable bowel syndrome	14	13	93
Vomiting	27	24	89
Emesis	60	52	87
Constipation	25	17	68

C Musculoskeletal

Indication	# Known	# Correct	% Correct
Akathisia	27	27	100
Arthritis	16	14	88
Dystonia	16	11	69
Osteoarthritis	12	7	58

molecule") and the comparison molecule, is converted into a bit string in four-atom increments across the molecule in a linear path (linear pattern) consisting of both bond and atom occurrence. The Tanimoto score is determined as:

$$T_S = N_C / (N_A + N_B - N_C)$$

where T_S is the Tanimoto Score, N_A and N_B are the number of bits set to 1 in the fingerprint of molecules A and B, respectively, and N_C is the total number of bits set to 1 found in the fingerprints of both ligands A and B. Two identical molecules will have a Tanimoto score of $T_S = 1$. We used a Tanimoto score of 0.6 or greater as an indicator of high similarity: we postulate that similar compounds could potentially have similar biological activity.

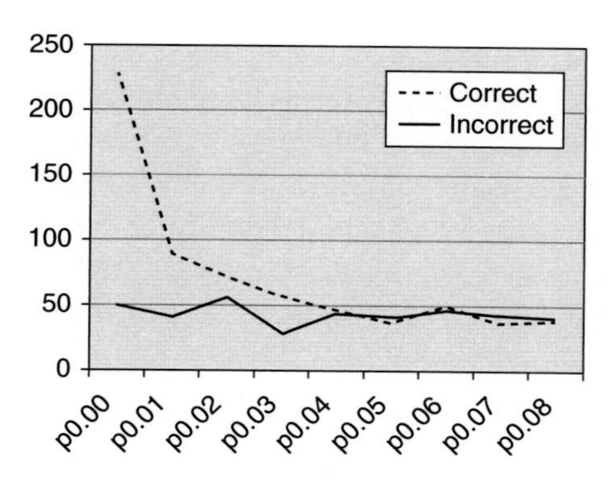

Figure 5.2 Plot of the number of predictions for both true and false positives and their associated significance coefficient (*p*-value). *p*-Values of less than 0.02 showed an exponential increase for the number of correct predictions.

Examples of learning from the CPA method

We determined the significance threshold for a protein vs. ontological term relationship using a null hypothesis test. For each ontological term, the level of occurrence was calculated using a binomial distribution test of success (Bernoulli experiment) across the entire phenotype ontology. The analysis showed that occurrences with a significance coefficient of 1% (p-value = 0.01, as based on a 2× sigma distribution) or greater were to be considered random (see Figure 5.2). Accordingly, we accepted any relationships reported with a p-value <0.01 as being a non-random association.

We analyzed our test set of 153 withdrawn drugs using the CPA method and for each obtained significant secondary pharmacology terms associated with them. These terms were then compared to the published adverse events (ADR) for each drug: on average, 65% of the cardiovascular, 74% of gastrointestinal, and 73% of the muscular group ADRs were identified correctly.

Cardiovascular results

When compounds that exhibit cardiovascular secondary pharmacology were reviewed for their success in identifying secondary effects ontologies (Table 5.2), our success rate was 93% for those with the tachycardia ADR; i.e., 46 drugs within our data set had this associated ADR and the CPA analysis correctly identified 43 of these compounds. Fifty-eight compounds associated with heart arrhythmia were correctly identified with a successful rate of 87%. Seven cardiovascular ontology terms had successful prediction rates of greater than 68% (Table 5.2A), the lowest of these being for "Myocardial infarction," where 13 of 19, or 68%, were predicted correctly. No correlation was found between the predicted secondary pharmacology and their representation in the ontology.

Gastrointestinal results

Table 5.2B notes five gastrointestinal ontology terms with correct prediction rates of 68% or greater. All nine small molecules in our data set that cause gastritis were properly identified. Equally successful, at 93%, was the ability of CPA to correctly identify those compounds which cause irritable bowel syndrome. Twenty-four of 27 compounds which cause vomiting were identified correctly. Our test set contained a total of 60 compounds that cause emesis, and CPA correctly identified 52, or 87%, of these.

Musculoskeletal results

Of the musculoskeletal ADR ontology terms, akathisia, or "restless leg" syndrome, had the highest percent correct with all 27 small molecules with this side effect identified. Three other ADR-associated ontology terms had correct prediction rates; i.e., 88% for arthritis, 69% correct for dystonia, and 58% correct for osteoarthritis (Table 5.2C).

CPA false positive rate

We determined the false positive rate using a test set of 120 compounds that do not have secondary pharmacology. This list contained a broad range of therapeutic compounds including antipsychotic drugs, such as benzisoxazole, to antihypertensive compounds, i.e., piperazines. CPA analysis showed false prediction rates of 19% across cardiovascular, gastrointestinal, and musculoskeletal group ADRs. Of the 120 compounds that were used in this analysis, 97 had no predicted ADRs with only 13 compounds identified falsely for a single phenotype ontology term. Nine of the 120 small molecules (7.5%) tested or were flagged falsely as having two or more predicted associated secondary pharmacology.

To address the correlation between phenotype ontologies and our screening database, we grouped together every ontology term to protein relationship observed using the 153 withdrawn/not approved compound test set and compared it to the overall frequency of occurrences within Medline. As ontology frequency is independent of observed protein/compound correlation, this demonstrates that the CPA approach relies upon a strong phenotype data set (Medline) for success, yet is not driven by such data in which high literature coverage exists.

Example: isoprenaline

Isoprenaline (CAS no. 7683-59-2) is a small molecule which targets both adrenergic receptors, beta-1 and beta-2 (ADRB1, ADRB2). Isoprenaline has been used in the treatment of asthma, anaphylaxis, bronchitis, emphysema, and other lung diseases to increase the airflow through the bronchial tubes. This compound has on-label ADRs that include increased heart rate and arterial pressure, which can lead to heart flutters. The CPA analysis indicates that the most significant phenotype associated with isoprenaline is tachycardia, with a *p*-value of 3.36e-06, nearly 3000× our cutoff significance. Isoprenaline exhibited inhibitory activity against several proteins (Buchheit *et al.*, 1985; Birrell *et al.*, 2002). These included alpha-1A/1B adrenergic receptors, beta-2 adrenergic receptor, D(2) dopamine receptor, and glucocorticoid receptor (GR). The number of publications listing "tachycardia" as well as each of these proteins is 106, 104, 296, 23, and 389, respectively.

Computational approaches expand knowledge of small-molecule effects

We describe a novel computational method for identifying potential secondary pharmacology associated with a compound. The method combines a simple compound similarity search with screening data and literature-based phenotype ontology. Using a test set of 153 small-molecule drugs withdrawn from the global markets, we demonstrate that the method correctly identifies an average of 65% of the cardiovascular, 74% of the gastrointestinal, and 73% of the musculoskeletal published secondary effects. We analyzed the most successful individual phenotype ontology terms, broken down by organ system grouping (Table 5.2) to assess reasons for the high correct prediction rates. For adverse cardiovascular symptoms, we reviewed the percent occurrence, within Medline, for the top seven phenotype ontology terms (Table 5.2A), in order to examine the correlation between CPA effectiveness and occurrence frequency in Medline. It could be theorized that a phenotype ontology term would have better probability for success within CPA if a higher percentage of Medline were covered by that term. However, after reviewing these data, no trend existed between the correct prediction rates and the term's background frequencies of occurrence within Medline. Similar findings were observed for both gastrointestinal and musculoskeletal phenotype ontologies. This is important because it shows the correct prediction rate for an ADR is not correlated to Medline ontology frequency.

CPA comparison to other approaches

Currently, several other computational approaches exist for *in silico* prediction of small-molecule effects on human physiology. Some computational tools, such as DEREK, are built on QSAR models of known carcinogenic compounds (Patlewicz *et al.*, 2004). Recent studies have shown that the predictive rates of DEREK and other commercially available programs, i.e., TOPKAT and CASETOX (Dearden, 2003; White *et al.*, 2003; Hillebrecht *et al.*, 2011; Krishna *et al.*, 2014), have been disappointing; with correct predictive genotoxicities rates (i.e., correlation with Ames assay results) of 21.3%–31.9% (Snyder *et al.*, 2004). DEREK and CPA do not yield directly comparable results because DEREK provides a probability of a compound being a mutagen. A major limitation of DEREK is that a novel or unique chemical series will yield a negative result. CPA uses a structure similarity search, which addresses this problem. CPA could be constructed to generate results similar to these genotoxitic predictive tools if ontology terms were generated that covered such events, e.g. "micronucleus" or "chromosomal lesion."

Other approaches such as novel chemogenomics techniques previously described (Niculescu *et al.*, 2004; Schuffenhauer and Jacoby, 2004) use systems biology to address secondary pharmacology occurrences. These novel and highly innovative *in silico* experiments are improving rapidly and current infrastructure needs for systems biology on such a scale are becoming available. The success for these and similar approaches remains high (Lipinski and Hopkins, 2004; Weinstein, 2005); however, such approaches should not solely rely on crude animal ED_{50}s (Mushal *et al.*, 2003) for their phenotype linkage. CPA ontology terms allows for a more precise assessment of secondary pharmacology than simple ED_{50} animal models. Specific identification of phenotype occurrences, such as tachycardia noted in the above-mentioned isoprenaline example, allows follow-up *in vivo* tests (e.g., cardiac tests) to be conducted, which address the flagged ontology term.

Distinct from predictive methods are databases containing experimentally determined biochemical-screening data across a large spectrum of targets, the BioPrint screening database being one example. BioPrint contains screening information for compounds across a complete matrix of assays. This data set includes over 1000 known drugs that are approved for human use, along with their associated FDA trial/on-label information including secondary pharmacology (ADRs) and the percentage of the population that exhibits these effects. Using biological activity spectra, we have previously shown (Fliri *et al.*, 2005) that small-molecule activity could be clearly joined to both structure and function. CPA and BioPrint analysis complement each other well, as the bulk, high-throughput analysis of CPA could be performed on a set of compounds followed up by a downstream biological activity spectra analysis using BioPrint. CPA gives primarily experimental-derived results, such as data obtained from screening databases, and would therefore complement the results obtained from a BioPrint analysis.

The CPA method takes into account not just information about the compound structure, but also (1) independent experimental information about how well a compound interacts with proteins, and (2) a protein's relationship to different phenotypes. It does not rely on publications about the compound itself (e.g., those discussing a drug's secondary pharmacology). Instead, it takes advantage of the experimental data where one or more similar compounds bind to one or more proteins, and those proteins have been associated with a given phenotype in a multiplicity of publications. The first part of the analysis, relating a compound to one or more targets, is useful in its own right because it provides a selectivity indication of off-target proteins. Similar useful information has been noted in such approaches using virtual screening methods not based solely on QSAR analysis (Chen and Ung, 2001).

Table 5.3 The most significant phenotype/toxicological terms and their associated *p* values for Isoprenaline (CAS 7683-59-2). The proteins that show interaction with isoprenaline are alpha-1A/1B adrenergic receptors, beta-2 adrenergic receptor, D(2) dopamine receptor and glucocorticoid receptor (GR).

Term	*p* Value
Tachycardia	3.36e-06
Hypotension	2.51e-05
Heart arrhythmia	5.88e-4
Congestive heart failure	5.72e-3
Bradycardia	1.02e-3

Supplemental Table 5.1 List of the 153 compound that were used for CPA benchmark analysis.

Name	CAS #
Acetanilide	103-84-4
Acetylsalicylic acid (pediatric)	50-78-2
Amfetamine	300-62-9
Aminoglutethimide	125-84-8
Aminophenazone	58-15-1
Aminophylline	317-34-0
Aminorex	2207-50-3
Amitriptyline	50-48-6
Amobarbital	57-43-2
Aprobarbital	77-02-1
Atropine in combination	51-55-8
Barbital	57-44-3
Benoxaprofen	51234-28-7
Benzyl alcohol	100-51-6
Berberine	2086-83-1
Bismuth subsalicylate	14882-18-9
Broxyquinoline (see also halogenated hydroxyquinoline derivatives)	521-74-4
Bucetin	1083-57-4
Buprenorphine	52485-79-7
Buspirone hydrochloride	33386-08-2
Canrenone	976-71-6
Cathine	492-39-7

Supplemental Table 5.1 (*cont.*)

Name	CAS #
Cefaloridine	50-59-9
Chenodeoxycholic acid	474-25-9
Chloramphenicol	56-75-7
Chloroquine	54-05-7
Chlorphentermine	461-78-9
Cinchophen	132-60-5
Ciprofibrate	52214-84-3
Cisapride	81098-60-4
Clemastine	15686-51-8
Clofibrate	637-07-0
Cloforex	14261-75-7
Clozapine	5786-21-0
Coumarin (synthetic)	91-64-5
Cyclandelate	456-59-7
Cyclopenthiazide	742-20-1
Cyproheptadine	129-03-3
Dantron	117-10-2
Depot medroxyprogesterone acetate (DMPA)	71-58-9
Dexamfetamine	51-64-9
Dexfenfluramine hydrochloride	3239-45-0
Dibenzepin hydrochloride	315-80-0
Dienestrol	84-17-3
Difenoxin	28782-42-5
Domperidone (injectable)	57808-66-9
Doxepin	1668-19-5
Droxicam	90101-16-9
Epinephrine	51-43-4
Ethylestrenol	965-90-2
Etomidate	33125-97-2
Fenfluramine	458-24-2
Fenoterol	13392-18-2
Flecainide	54143-55-4
Floctafenine	23779-99-9
Flunitrazepam	1622-62-4
Fluoxetine	54910-89-3
Fluvoxamine	54739-18-3
Furazolidone	67-45-8
Gemfibrozil	25812-30-0
Grepafloxacin hydrochloride	161967-81-3

Supplemental Table 5.1 (*cont.*)

Name	CAS #
Griseofulvin	126-07-8
Halogenated hydroxyquinoline derivatives	148-24-3
Hexobarbital	56-29-1
Ibuprofen	15687-27-1
Indometacin and indometacin farnesil	53-86-1
Isocarboxazid	59-63-2
Isoprenaline	7683-59-2
Isoxicam	34552-84-6
Ketorolac	74103-06-3
Lamotrigine	84057-84-1
Levamfetamine	156-34-3
Levamisole hydrochloride	16595-80-5
Levarterenol	51-41-2
Loperamide	53179-11-6
Loxoprofen sodium	68767-14-6
L-Tryptophan	73-22-3
Mazindol	22232-71-9
Mefloquine	53230-10-7
Megestrol acetate	3562-63-8
Melatonin	73-31-4
Mephenesin	59-47-2
Mesna	19767-45-4
Methamphetamine	537-46-2
Methylphenidate	113-45-1
Metoclopramide (pediatric)	364-62-5
Mianserin	24219-97-4
Minocycline	10118-90-8
Mofebutazone	2210-63-1
Nialamide	51-12-7
Nifedipine	21829-25-4
Nitrofurantoin	67-20-9
Noscapine	128-62-1
Oxyphenisatine acetate	115-33-3
Paracetamol	103-90-2
Pemoline	2152-34-3
Pentazocine	359-83-1
Pentobarbital	76-74-4

Supplemental Table 5.1 (*cont.*)

Name	CAS #
Phenazone	60-80-0
Phenazopyridine	94-78-0
Phendimetrazine	634-03-7
Phenobarbital	50-06-6
Phenol	108-95-2
Phenolphthalein	77-09-8
Phenoxybenzamine	59-96-1
Phentermine	122-09-8
Phentolamine mesilate	65-28-1
Phenylpropanolamine	14838-15-4
Pipamazine	84-04-8
Potassium canrenoate	04-06-2181
Practolol	6673-35-4
Promethazine	60-87-7
Propafenone	54063-53-5
Propofol	2078-54-8
Pyridoxine (Vitamin B6)	65-23-6
Pyritinol	1098-97-1
Quinine sulfate	804-63-7
Remifentanil	132875-61-7
Remoxipride	117591-79-4
Scopolamine	51-34-3
Secobarbital	76-73-3
Sertindole	106516-24-9
Sotalol	3930-20-9
Sparfloxacin	110871-86-8
Spironolactone	52-01-7
Strychnine and salts	57-24-9
Sulfacarbamide	547-44-4
Sulfadicramide	115-68-4
Sulfadimidine	57-68-1
Sulfamerazine sodium	127-58-2
Sulfanilamide	63-74-1
Sulfathiazole	72-14-0
Sulfisomidine	515-64-0
Sultopride	53583-79-2
Sumatriptan	103628-48-4
Suprofen	40828-46-4

Supplemental Table 5.1 (*cont.*)

Name	CAS #
Suxamethonium chloride	71-27-2
Temazepam	846-50-4
Terodiline	15793-40-5
Testosterone propionate (injectable)	57-85-2
Thalidomide	50-35-1
Tianeptine sodium	30123-17-2
Tiaprofenic acid	33005-95-7
Tocainide	41708-72-9
Tramadol	27203-92-5
Trazodone	19794-93-5
Triazolam	28911-01-5
Trimipramine	739-71-9
Vinbarbital	125-42-8
Vincamine	1617-90-9
Warfarin	81-81-2
Zimeldine	56775-88-3
Zopiclone	43200-80-2

Supplemental Table 5.2 List of the 120 compound that were used for CPA false positive analysis.

Name	CAS #
Aceclofenac	89796-99-6
Actarit	18699-02-0
Alfuzosin HCl	81403-68-1
Amantadine HCl	665-66-7
Aminophylline ethylenediamine salt	317-34-0
Amlodipine maleate salt	88150-42-9
Amoxapine	14028-44-5
Ampiroxicam	99464-64-9
Aspirin	50-78-2
Benzonatate	104-31-4
Bepridil HCl	68099-86-5
Betamethasone dipropionate	5593-20-4
Betaxolol HCl	63659-19-8
Bisoprolol	66722-44-9
Bumetanide	28395-03-1

Supplemental Table 5.2 (*cont.*)

Name	CAS #
Buprenorphine HCl	53152-21-9
Bupropion HCl	31677-93-7
Captopril	62571-86-2
Carvedilol	72956-09-3
Cefpodoxime proxetil	87239-81-4
Cefprozil Z-isomer	92665-29-7
Ceftriaxone sodium E-isomer	74578-69-1
Cerivastatin	145599-86-6
Cetraxate HCl	27724-96-5
Chlormadinone acetate	302-22-7
Cinoxacin	28657-80-9
Citalopram HBr	59729-32-7
Clenbuterol HCl	21898-19-1
Clindamycin phosphate	24729-96-2
Clomipramine	303-49-1
Clonazepam	1622-61-3
Clonidine HCl	4205-91-8
Cyclobenzaprine HCl	6202-23-9
Desipramine HCl	58-28-6
Digitoxin	71-63-6
Dihydroergotamine mesylate	6190-39-2
Dinoprost tromethamine	38562-01-5
Dirithromycin	62013-04-1
Donepezil HCl	120011-70-3
Enalapril maleate	76095-16-4
Enalaprilat	76420-72-9
Enoxacin	74011-58-8
Esomeprazole magnesium	73590-58-6
Ethacrynic acid	58-54-8
Fenoldopam	67227-56-9
Fexofenadine HCl	138452-21-8
Fleroxacin	79660-72-3
Fludrocortisone acetate	514-36-3
Fluoxetine HCl	56296-78-7
Fluvoxamine maleate	61718-82-9
Fosinopril	98048-97-6
Galantamine HBr	1953-04-4
Gatifloxacin	112811-59-3

Supplemental Table 5.2 (*cont.*)

Name	CAS #
Gemfibrozil	25812-30-0
Ibudilast	50847-11-5
Irinotecan HCl	100286-90-6
Irsogladine maleate	84504-69-8
Itopride	122898-67-3
Lamotrigine	84057-84-1
Levofloxacin	100986-85-4
Lincomycin HCl	859-18-7
Lisinopril	76547-98-3
Lovastatin	75330-75-5
Maprotiline HCl	10347-81-6
Mefloquine HCl	51773-92-3
Memantine HCl	41100-52-1
Mepenzolate bromide	76-90-4
Methotrexate	59-05-2
Metoclopramide HCl	54143-57-6
Midazolam	59467-94-6
Mirtazapine	61337-67-5
Monoprilat	
Nalbuphine HCl	59052-16-3
Naratriptan	121679-13-8
Nelfinavir	159989-64-7
Neostigmine bromide	114-80-7
Nevirapine	129618-40-2
Nicardipine HCl	55985-32-5
Nimodipine	66085-59-4
Nitrendipine	39562-70-4
Norfloxacin	70458-96-7
Octreotide acetate	79517-01-4
Olanzapine	132539-06-1
Paroxetine	61869-08-7
Perindopril TFA salt	
Physostigmine salicylate	57-64-7
Procainamide HCl	614-39-1
Protriptyline HCl	1225-55-4
Quinapril	85441-61-8
Ramipril	87333-19-5
Rimantadine HCl	1501-84-4

Supplemental Table 5.2 (*cont.*)

Name	CAS #
Risperidone	106266-06-2
Ropivacaine	84057-95-4
Rosuvastatin Ca	287714-41-4
Sertraline	79617-96-2
Succinylcholine chloride	71-27-2
Sulfamethoxazole	723-46-6
Sulfasalazine	599-79-1
Sumatriptan	103628-46-2
Tacrine	321-64-2
Tacrolimus	109581-93-3
Telmisartan	144701-48-4
Temazepam	846-50-4
Terazosin HCl	63074-08-8
Terfenadine	50679-08-8
Theophylline	58-55-9
Tiagabine HCl	115103-54-3
Ticlopidine HCl	53885-35-1
Tizanidine HCl	51322-75-9
Tocainide HCl	35891-93-1
Tosufloxacin tosylate	100490-94-6
Tranilast	53902-12-8
Triazolam	28911-01-5
Trimebutine	39133-31-8
Troxipide	30751-05-4
Valdecoxib	
Venlafaxine HCl	99300-78-4
Voriconazole	N/F
Zidovudine	30516-87-1
Ziprasidone	none

Computational methods like the CPA are relatively inexpensive and high-throughput when compared to *in vivo* or *in vitro* safety models and can be used in the early stages of drug discovery (see, for example, Table 5.3). Large libraries of compounds can be quickly interrogated for possible secondary pharmacology even before a chemist steps into a laboratory. Alternatively, drug discovery programs may have a number of chemically distinct series of compounds for a single target: these can be easily triaged for possible safety liabilities. Those compounds identified as having few potential liabilities could then be quickly placed into additional cell-based *in vitro* assays, thereby coupling *in silico* and *in vitro* approaches in a safety paradigm that increases the chances for NCE approval while addressing potential problems in the early stages of discovery.

References

Arrowsmith, J. Trial watch: Phase II failures: 2008–2010. *Nature Reviews Drug Discovery*. 2011a;10:328–329.

Arrowsmith, J. Trial watch: Phase III and submission failures: 2007–2010. *Nature Reviews Drug Discovery*. 2011b;10:87.

Baker, N. C. and Hemminger, B. M. Mining connections between chemicals, proteins and diseases extracted from Medline annotations. *Journal of Biomedical Informatics*. 2010;43:510–519.

Birrell, M., Crispino, N., Hele, D. J., *et al*. Effect of dopamine receptor agonists on sensory nerve activity: Possible therapeutic targets for the treatment of asthma and COPD. *British Journal of Pharmacology*. 2002;136:620–628.

Buchheit, K. H., Engel, G., Mutschler, E. and Richardson, B. Study of the contractile effect of 5-hydroxytryptamine (5-HT) in the isolated longitudinal muscle strip from guinea-pig ileum. Evidence for two distinct release mechanisms. *Naunyn Schmiedeberg's Archives of Pharmacology*. 1985;329(1):36–41.

Campbell, S. J., Gaulton, A., Marshall, J., *et al*. Visualizing the drug target landscape. *Drug Discovery Today*. 2010;15(1–2):3–15.

Chen, Y. and Ung, C. Prediction of potential toxicity and side effect protein targets of a small molecule by a ligand–protein inverse docking approach. *Journal of Molecular Graphics and Modeling*. 2001:20:199–218.

DataMonitor – Pharmaceutical Benchmark Report. www.datamonitor.com.

Dearden, J. C. *In silico* prediction of drug toxicity. *Journal of Computer Aided Molecular Design*. 2003;17:119–127.

Fliri, A., Loging, W., Thadeio, P. and Volkmann, R. Biological spectra analysis: Linking biological activity profiles to molecular structure. *Proceedings of the National Academy of Sciences, USA*. 2005;102:261–266.

Greene, N. Computer systems for the prediction of toxicity: An update. *Advances in Drug Delivery Review*. 2002;31:417–431.

Hillebrecht, A., Muster, W., Brigo, A., *et al*. Comparative evaluation of in silico systems for Ames test mutagenicity prediction: scope and limitations. *Chemical Research in Toxicology*. 2011;24:843–854.

Krejsa, C. M., Horvath, D., Rogalski, S. L., *et al.* Predicting ADME properties and side effects: The BioPrint approach. *Current Opinion in Drug Discovery Development*. 2003;6:470–480.

Krishna, K. A., Saryu, G. and Krishna, G. SAR genotoxicity and tumorigenicity predictions for 2-MI and 4-MI using multiple SAR software. *Toxicological Mechanisms and Methods*. 2014;24:284–293.

Leishman, D. J. and Rankovic, Z. Drug discovery vs hERG. In *Tactics in Contemporary Drug Discovery* (pp. 225–260), ed. N. A. Meanwell. Berlin, Springer, 2014.

Lipinski, C. A. and Hopkins A. Navigating chemical space for biology and medicine. *Nature Insight*. 2004;432:855–861.

Lipinski, C. A., Lombardo, F., Dominy, B. W. and Feeney, P. J. Experimental and computational approaches to estimate solubility and permeability in drug discovery and development settings. *Advanced Drug Delivery Reviews*. 1996;23:3–25.

Loging, W. L., Rodriguez-Esteban, R., Hill, J., Freeman, T. and Miglietta, J. Chemoinformatic/bioinformatics analysis of large corporate databases: application to drug repurposing. *Drug Discovery Today: Therapeutic Strategies*. 2011;8:109–116.

Munos, B. Lessons from 60 years of pharmaceutical innovation. *Nature Reviews Drug Discovery*. 2009; 8:959–968.

Mushal, S. M., Jha, S. K., Kishore, M. P. and Tyagi, P. A simple and readily integratable approach to toxicity prediction. *Journal of Chemical Information and Computer Sciences*. 2003;43:1673–1678.

Niculescu, S. P., Atkinson, A., Hammond, G. and Lewis, M. Using fragment chemistry data mining and probabilistic neural networks in screening chemicals for acute toxicity to the fathead minnow. *SAR QSAR Environmental Research*. 2004;15:293–309.

Patlewicz, G. Y., Rodford, R. and Walker, J. D. Quantitative structure–activity relationships for predicting mutagenicity and carcinogenicity. *Environmental and Toxicological Chemistry*. 2003;22:1885–1893.

Patlewicz, G. Y., Basketter, D. A., Pease, C. K., *et al*. Further evaluation of quantitative structure–activity relationship models for the prediction of the skin sensitization potency of selected fragrance allergens. *Contact Dermatitis*. 2004;50:91–97.

Sanderson, D. and Earnshaw, C. Computer prediction of possible toxic action from chemical structure; The DEREK system. *Human Experimental Toxicology*. 1991;10:261–273.

Schuffenhauer, A. and Jacoby, E. Annotating and mining the ligand–target chemogenomics knowledge space. *Drug Discovery Today: BioSilico*. 2004;2:190–200.

Senese, C. L., Duca, J., Pan, D., Hopfinger, A. J. and Tseng, Y. J. 4D-Fingerprints, universal QSAR and QSPR descriptors. *Journal of Chemical Information and Computer Sciences*. 2004;27:1526–1539.

Snyder, R. D., Pearl, G. S., Mandakas, G., *et al*. Assessment of the sensitivity of the computational programs DEREK, TOPKAT, and MCASE in the prediction of the genotoxicity of pharmaceutical molecules. *Environmental and Molecular Mutagenesis*. 2004;43:143–158.

Thomas, D., Karle, C. A. and Kiehn, J. Modulation of HERG potassium channel function by drug action. *Annals of Medicine*. 2004;36: 41–46.

United Nations, Department of Economic and Social Affairs. *Consolidated List of Products Whose Consumption and/or Sale Have Been Banned, Withdrawn, Severely Restricted or not Approved by Governments* 8th Issue (2003).

Villoutreix, B. O. and Taboureau, O. Computational investigations of hERG channel blockers: new insights and current predictive models. *Advanced Drug Delivery Reviews*. 2015;68:72–82.

Waring, M. J., Arrowsmith, J., Leach, A. R., *et al*. An analysis of the attrition of drug candidates from four major pharmaceutical companies. *Nature Reviews Drug Discovery*. 2015;14:475–486.

Weinstein, J. N. Linking drugs and genes: Pharmacogenomics, pharmacoproteomics, bioinformatics, and the NCI-60. In *Oncogenomics: Molecular Approaches to Cancer*, ed. C. Brenner and D. Duggan. Hoboken, NJ: John Wiley & Sons, 2005.

White, A. C., Mueller, R. A., Gallavan, R. H., Aaron, S. and Wilson, A. G. A multiple in silico program approach for the prediction of mutagenicity from chemical structure. *Mutation Research*. 2003;5:77–89.
Willett, P., Barnard, J. and Downs, G. Chemical similarity searching. *Journal of Chemical Information and Computing Science*. 1998;38:983–996.

6 Data visualization and the DDP process

Ke Xu

Data visualization denotes the techniques of visually presenting complex data sets to achieve goals such as displaying multiple data dimensions simultaneously, connecting related data points from data sets, or showing data distribution patterns. They are of great value for data processing, data analysis, and data presentation activities.

Genomics and functional genomics are the major driving forces for the development and utilization of visualization tools in biological fields. Following the completion of genomic sequencing projects of human and other model organisms around the beginning of this century, our knowledge of genes has jumped to the tens of thousands per species. Expression profiling microarray can generate millions of data points per experiment. The challenge of the huge data set size and the need to integrate different data sources in analyses prompted significant research and development work by both academic and industrial bioinformaticians. As a result, many visualization methods, proposals, and tools for biological data have been developed thus far. This chapter will describe the problems and solutions for the visualization of three basic and largest (thus, most challenging) genomics/functional genomics data types. More specifically, the first two sections will discuss visualization of sequence data and pathway/gene network data, which are two data types specific to genomics and other biology fields. In the third section, we will review visualization methods of numeric data, such as expression profiling data, proteomic data, and genotyping data. Most of the techniques in the section can also be applied to other areas. However, some topics, such as viewing numeric data in the context of genome or pathways, are still biology-specific.

Sequence and genomes

The genome is the complete set of genetic materials for an organism, which includes genes, regulatory and replication-related sequences, as well as non-functional intergenic regions. For most organisms other than RNA viruses, long linear or circular DNA molecules form the biochemical basis of the genome that stores all the

Bioinformatics and Computational Biology in Drug Discovery and Development, ed. W.T. Loging. Published by Cambridge University Press.

genetic information. Visualization of the genome refers to the visual display of the DNA sequences and associated annotations. Depending on the visualization purposes, genome visualization tools can be classified into two categories: sequence viewer for visualizing sequence and annotations, and genome alignment viewer, for comparing different genomes.

Visualizing genomic sequence

Other than the sequence itself, which is a string of A, G, C, and T characters, annotation data are another important part of genome data. Annotations are information associated with a segment or a group of segments on the sequence. They could be biological features identified from analysis of the sequence, or information about the sequencing process, such as ambiguity bases, border of contigs, etc. Some of the common examples of biological features are: location of genes, exons, 5′ or 3′ untranslated regions, promoter elements, transcription factor binding sites, SNPs, etc. A more complete list of features can be found in the Genbank/EMBL/DDBJ feature table definition document (www.ncbi.nlm.nih.gov/collab/FT/). Also related to annotation data are computational analysis results that can be mapped to the genome, such as BLAST alignment of EST sequences to the genome or exons predicted by the Genscan program. These are the data contents used by curators to generate annotations. However, in some cases where no other information is available, these analysis results could be used directly as the annotations of the genome. In fact, some tools do not differentiate these two types and treat them together as annotations.

In the early days before genomic sequencing, the standard format for storing and exchanging sequence information was Genbank/EMBL/DDBJ format, in which each entry contains both sequence and annotations, with annotations organized in the so-called feature tables. This format is no longer a desired solution for whole-genome sequences because the size of each sequence (up to a chromosome) is often too large to be stored as one entry. As a result, many alternative solutions for data storage and exchange were developed. One solution is directly exchanging and distributing relational databases of whole genomes. In this format, annotation data are stored in database tables while sequence data could be either in tables or stored in simple sequence files such as FASTA files. For annotation data, there is another popular exchange and storage format, the so-called GFF file format (www.sanger.ac.uk/Software/formats/GFF/GFF_Spec.shtml). GFF stands for generic feature format. It is a tab-delimited text file in which each row is a feature (one data point of annotations) and columns are assigned to data content such as feature name, start and stop location, strand, etc. Because there is no sequence data in GFF files, these files are normally used together with corresponding FASTA sequence files. Lastly, a number of XML formats have been developed for genomic sequence and annotation data, e.g., Chaodo XML formats for fruit fly genomic data.

Just as various formats for sequence and annotation data storage exist for different sequence types, different genome visualization tools were developed for different uses. On the one end, database annotators who generate annotation data need fast tools with fully interactive controls and write access to the underlying database. Java applications like Artemis and Apollo were developed for this purpose (Lewis *et al.*, 2002; Berriman and Rutherford, 2003). These tools normally can read data from all the above data formats. On the other hand, general and occasional users mainly need maintenance free, read-only access to centralized databases. Web interface with Java scripts works best for these requirements. ENSEMBL (www.ensembl.org) and UCSC genome browser (http://genome.ucsc.edu/browser) are two most popular tools in this category.

Despite the differences in platforms, software architectures and user bases, all these tools use the same general concept to visualize the genomic sequence and annotation data, which is to display the genome horizontally, and align annotation data and computational analysis results to the matching genomic locations. Depending on zooming factors, sequence and annotation data can be displayed as graphic elements (blocks or arrows) or real sequences. With this layout, users can easily identify annotations associated with a particular genomic region, compare different computational analysis results of that region, and examine details at the sequence level when needed.

Because of the huge size of chromosomes, navigation and zooming become important features for users to locate the regions of interest. All these tools provided controls to zoom in and zoom out of the region displayed, and a concept of "semantic zooming" is adopted. Semantic zooming means the system would display the data in the form that is appropriate to the zooming factor. When a large genomic region is displayed, a box or similar graphic elements would be used to represent the sequence. However, if the region is small enough, the actual sequence would be displayed. To locate and position a region, multiple options are often provided. One is through text-based input such as gene name, chromosome, and base pair number. Another more flexible option is "drag and select," in which users select a region to be displayed by highlighting with the computer mouse. The "drag and select" feature of ENSEMBL is particularly useful because of its three-view display. The three views are three sequence viewing windows from top to bottom showing the chromosome, a large overview genomic region, and a smaller detailed region. ENSEMBL uses these three views to display the same target genomic region at three different zoom levels. Users can drag and select in the upper view to select the region to be displayed in the lower views. Drag and select an area in the bottom detailed view would change the view itself into displaying the selected area only (Figure 6.1).

The amount of annotation and computational analysis results is another challenging problem for visualizing genomic data. For example, some genes might have hundreds of expressed sequence tags (ESTs) matching part of the gene from the EST database, which makes the effort of displaying them quite difficult. Most tools address this problem by organizing annotation data into groups and providing options to toggle the display mode for each group. Users can choose to hide a group,

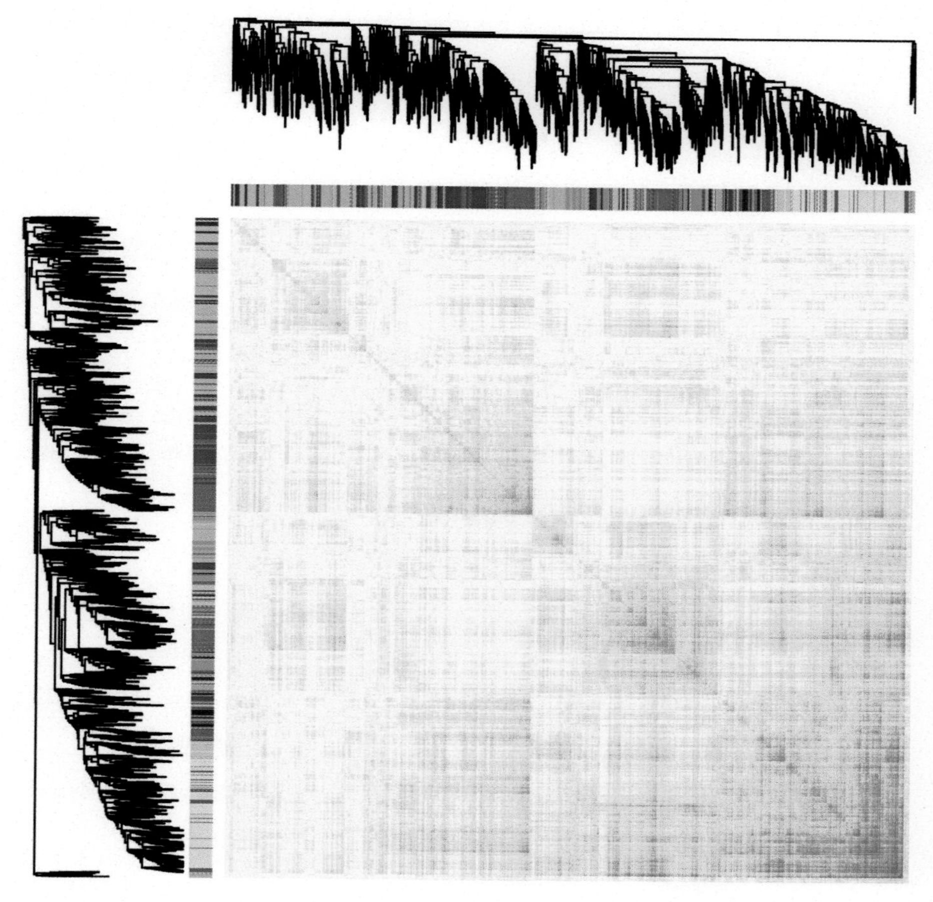

Figure 2.4 Topological overlap map. This is an example of a topological overlap map (TOM) constructed from weighted pairwise calculation of gene expression correlation between each transcript and every other transcript across the study population. Transcripts are arrayed across the x-axis and y-axis; each cell is a weighted correlation between the two corresponding transcripts. Hierarchical clustering is performed along each axis, resulting in the generation of highly interconnected modules of transcripts along the diagonal. These modules represent biologically relevant functional units related to the phenotype of the study subjects.

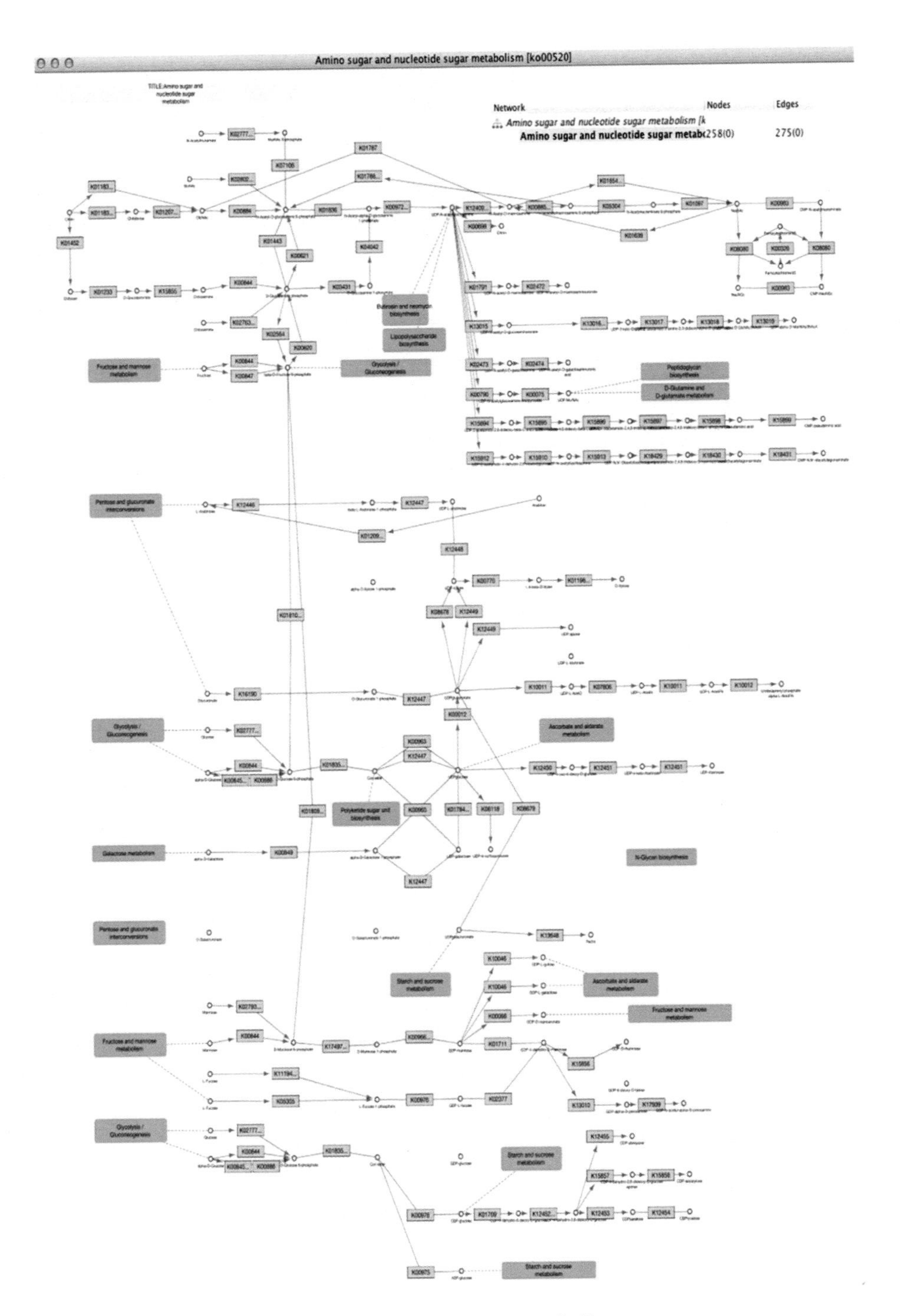

Figure 4.3 Example of the amino sugar and nucleotide sugar metabolism.

B-cell activating factor (BAFF)

Tissue	uHits
B-Lymphocytes	**641**
Spleen	**156**
Lymphoid Tissue	**118**
Bone Marrow	**116**
Salivary Glands	79
Lymphocytes	**58**
Appendix	**54**
Macrophages	**54**
Monocytes	**54**
Lung	49
Blood	**44**
Dendritic Cells	**44**
Germinal Center	**44**
T-Lymphocytes	**32**
Serum	**20**
Palatine Tonsil	**18**
Plasma Cells	**18**
Bone Marrow Cells	**16**
Leukocytes	**16**
Endothelium	14

Fractalkine Receptor (GPR13)

Tissue	uHits
Brain	**323**
Neurons	**259**
Astrocytes	**243**
Macrophages	**204**
Spinal Cord	**195**
Neuroepithelial Cells	**142**
Microglia	**117**
Endothelial Cells	**122**
Leukocytes	**103**
CD4-Positive T-Lymphocytes	**102**
Kidney	101
Dendritic Cells	**98**
Lung	91
Spleen	**90**
Neuroglia	**85**
Neural Stem Cells	**71**
Endothelium	58
Liver	57
Breast	52
Heart	51

Figure 4.8 Target localization within tissues by semantic relationships using natural language processing. Specific proteins and their synonyms were queried for relationships connected with laboratory methodologies used to directly observe the presence or absence of target expression in tissues and cells. Unique hits (uHits) were then curated, summed and ranked by frequency of uHits to derive a consensus tissue of interest. Bolded results are expected biological sites of expression for a particular drug target. Bolded red results are potential "off-target" locations indicating a tissue or cell type that is not within the scope of the intended pharmacological mechanism of a targeting drug concept.

Figure 6.2 Several visualizations for sequence comparisons: (a) alignment, (b) dotplot, (c) phylogenetic tree, and (d) sequence logo.

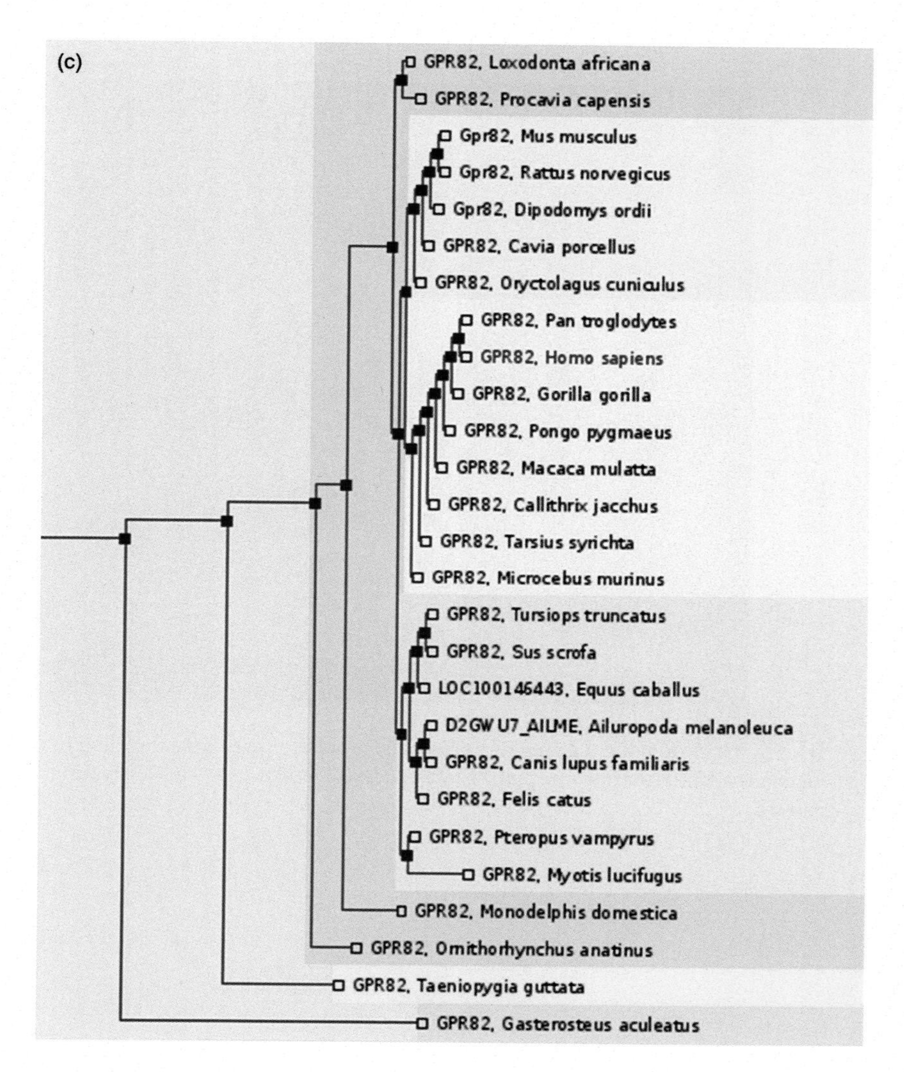

Figure 6.2 (*cont.*)

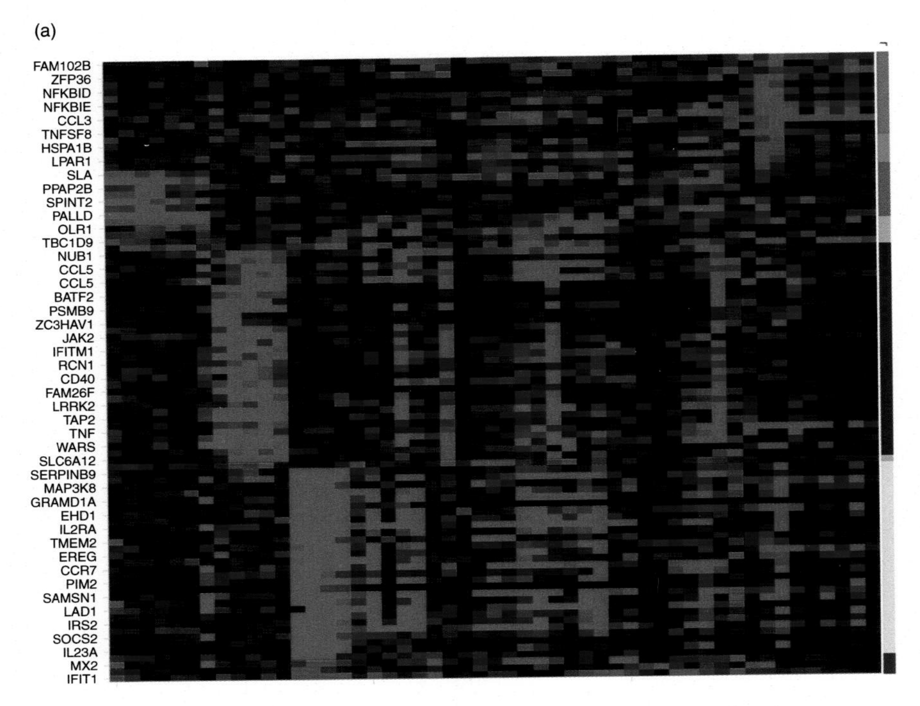

Figure 6.9 Heat map (a) showing a group of genes with expression pattern correlated to cancer metastasis status and profile map (b) showing clusters of genes with circadian expression pattern.

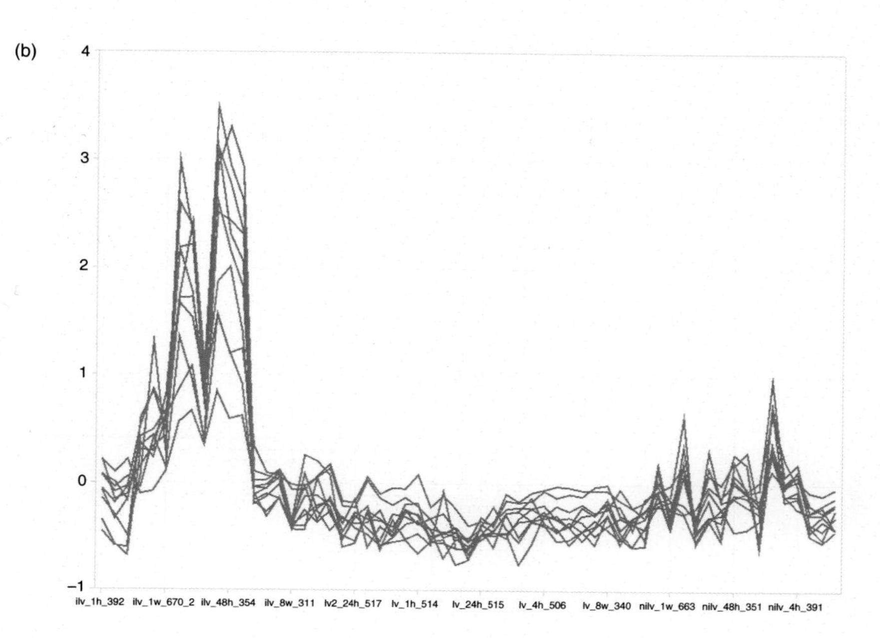

Figure 6.9 (*cont.*)

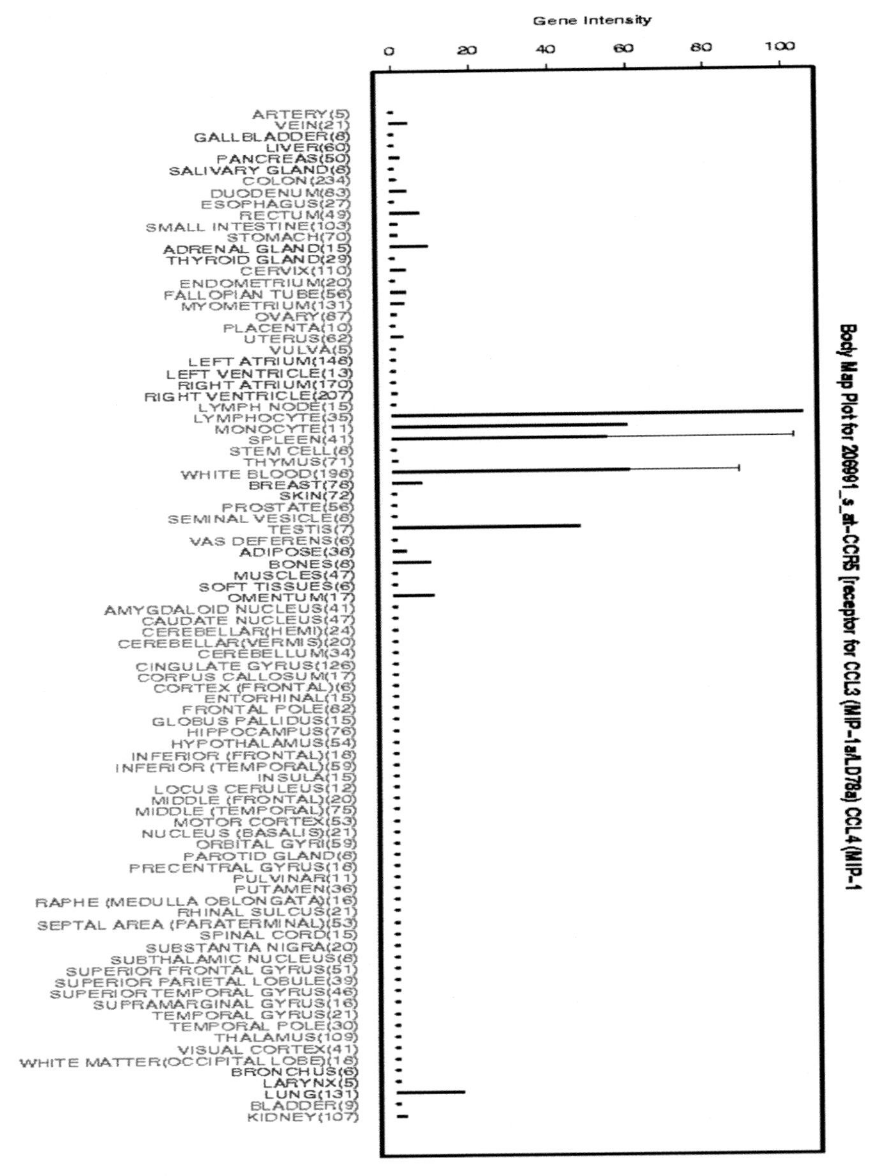

Figure 8.5 Body map showing expression of CCR5 in different body compartments and organs. Body map gene expression analysis conducted using Affymetrix GeneChip® technology in order to identify CCR5 expression across more than 2,000 tissue samples obtained using 90 whole pathologically normal human tissues from 13 different organ systems. The analysis showed the highest expression of CCR5 was on lymphocytes, monocytes, and spleen tissue, with no detectable expression of CCR5 in the liver (CCR5 gene expression based on an analysis of 60 normal samples of human liver tissue derived from both male and female subjects). Numbers in parentheses represent the number of tissue samples analyzed for each body compartment/organ.

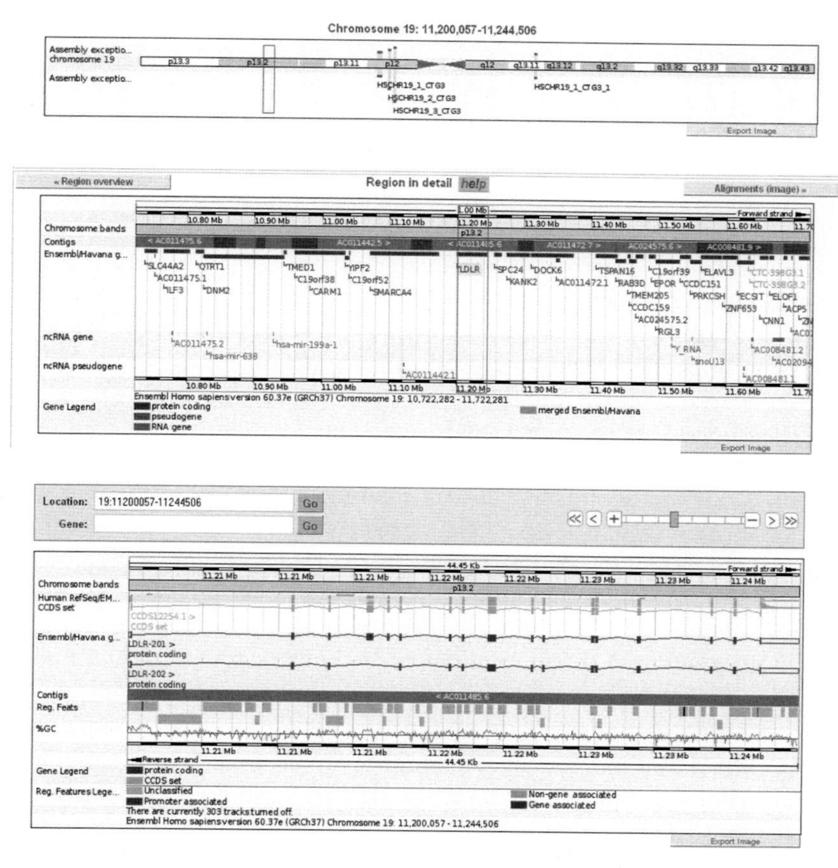

Figure 6.1 ENSEMBL uses three windows to display the genome. The top window shows the chromosome with a red box highlighting the region to be shown in the region view window. The middle window is the region view window. It displays genes in this region as red bars. Again, a red box highlights the subregion to be displayed in the detail window at the bottom. The detail window is used for displaying annotations of that subregion, such as predicted exons, matching mRNA sequences.

or display a group in dense mode or full mode. Dense mode combines annotation data of the same group into one track and provides users with a good compromise between not showing and showing all the details of the data. Normally, an annotation track just tells users the location and the type of annotation data. This can be done with a box aligned to the matching genomic location, or boxes linked by lines to show matching sections spaced by intronic or intergenic regions. However, some data types have numerical values along the sequence, such as GC content, homology conservation scores to other species. These can be displayed as small histograms along the track. Because there are two strands of the DNA sequence and most annotation data are specific to one strand, most tools have adopted specific notations to represent the strand information. Some tools draw annotations as an arrow to indicate the annotation data are on the forward or reverse strand. Other tools choose to display the genome in the center of the view and display annotation data for one strand above the genome and annotation data for the other strand below the genome. For these tools, a very convenient feature is to be able to reverse the genomic DNA with one click so the appropriate direction is displayed.

Visualizing sequence comparisons

The goal of sequence comparison is to identify and measure the similarity between two or more related sequences after their separation in the evolution process. It is an important technique in identifying genes or regulatory elements in newly sequenced genomes, predicting the structure and function of novel genes based on their homology to known genes, and studying evolutionary relationships between different genes or organisms. The key of sequence comparison is the identification of maximum alignment between the sequences to be compared. In this process, identical or similar residues are identified and the sequences are aligned according to these residues. Usually, gaps are introduced at the locations along the sequences where deletions or insertions have happened. With correct alignment, similarity measurements such as percent identity or evolutionary distances can then be calculated and further data-mining processes, such as pattern/motif discovery, can be performed.

Several methods have been widely used for visualizing sequence comparisons (Figure 6.2). First, sequence alignment can be used directly to show the conserved and changed regions between the sequences. Various visual elements, such as colors, shadings, and boxes, are used to highlight conserved regions or amino acid properties. Some visualization tools even provide the integration of other information of each sequence (such as transmembrane domains from the sequence feature table) in the view. Compared to other techniques, sequence alignment is good for displaying multiple sequences and it is easy to show conserved motifs or active sites in the alignment. Dot plot is another technique that presents the similarities without the need to first generate full alignment of two sequences. In this plot, comparisons are made on all combinations of short subsequences between two sequences and results are plotted with the x-axis representing one sequence and the y-axis representing the other. A dot represents the similarity between two subsequences at the plotted location being higher than a threshold. Thus, a pair of homologous sequences would generate a dot plot of a diagonal line. Dot plot is particularly useful when the two sequences are very large in length. However, it cannot be used when more than two sequences are to be compared. In addition to these two methods, phylogenetic tree and sequence logo are two methods that can be used to show the results of sequence comparisons. Phylogenetic trees show the relative distance of sequences to be compared, while sequence logos show the consensus sequence of the group of sequences.

With the completion of sequencing of multiple genomes, it is now of great interest to compare long genomic regions containing many genes between different

(b) Dotmatcher: genomic·human4 vs genomic·monkey3

(windowsize = 20, threshold = 65.00 30/09/08)

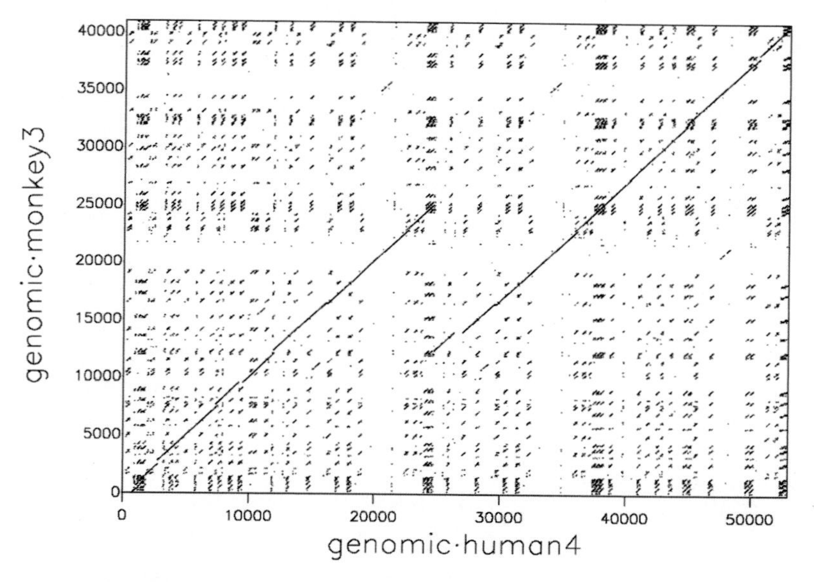

Figure 6.2 Several visualizations for sequence comparisons: (a) alignment, (b) dot plot, (c) phylogenetic tree, and (d) sequence logo. A black and white version of this figure will appear in some formats. For the color version, please refer to the color plate section.

species. Two families of software tools, VISTA and PipMaker, have emerged as the tools of choice for this purpose (Schwartz *et al.*, 2000; Frazer *et al.*, 2004). Both tools have visualization layouts similar to the genome browser which can display features like chromosome location, gene and exon location, etc. Added in these tools are the graphic elements showing the similarity between the query genomes (Figure 6.3). In VISTA tools, a curve is used to show the similarity of a running window along the aligned genomes. In PipMaker, the similarity in terms of percent identity (and hence the term Pip, for percent identity plot) is represented by dots. Again, color shadings are used in both tools for highlighting conserved regions.

When the size of the query genomic regions grows even larger, it is quite common that subregions of the genomes may be reversed or deleted, and unrelated regions may be inserted in the region. To show this type of homology (synteny),

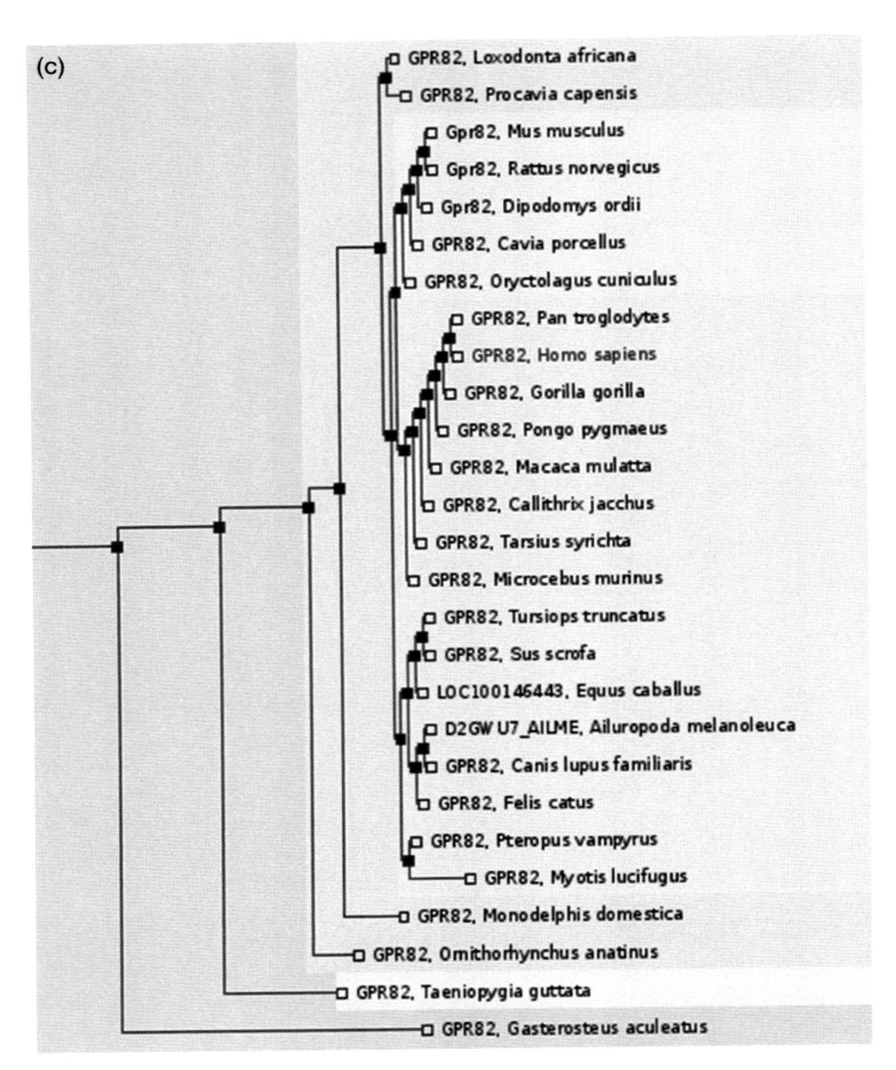

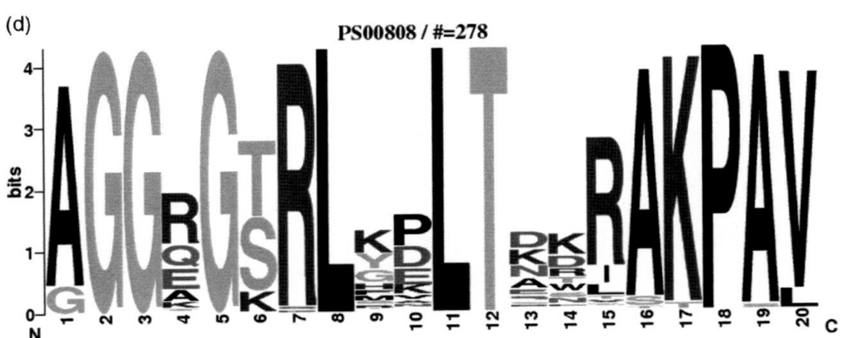

Figure 6.2 *(cont.)*

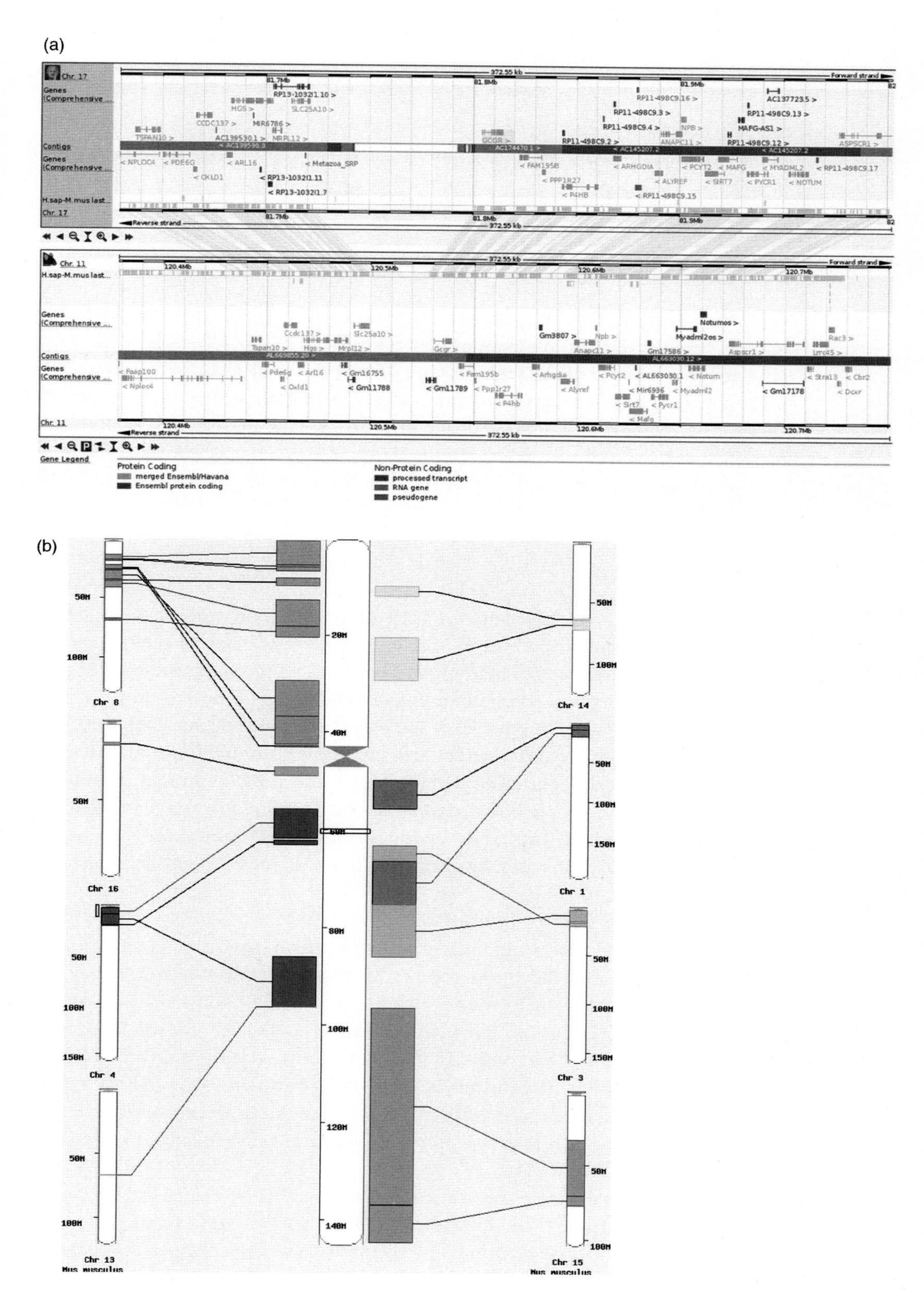

Figure 6.3 ENSEMBL comparative genomics region view (a) and synteny view (b).

more visual aid is needed to help users identify matching regions and their directions. Apollo's synteny view is a good example of this type of visualization (Figure 6.3b).

Pathways and molecular network data

With the data generated by functional genomics experiments and literature data mining, network data of interactions between various bio-molecules have been rapidly growing in recent years. Pathways are one type of molecular network in which nodes are gene products or metabolites, edges are biochemical or biophysical processes, and the networks represent one or more specific biological functions. For historical reasons and due to the unique characteristics of pathway data, they are normally visualized with tools different from the tools handling regular network data. This section will discuss the visualization of these network data types.

Visualizing pathways

Biological pathways are typically grouped into two broad categories: metabolic pathways, and signaling pathways. Metabolic pathways describe the network of biochemical reactions occurring in organisms that lead to the synthesis or degradation of various organic compounds (metabolites). Signaling pathways refer to pathways that represent the passage of biological signals through a cascade of biochemical reactions or physical interactions, such as receptor binding and activation, phosphorylation and dephosphorylation of proteins, protein–protein interaction, and DNA–protein interaction. Depending on the downstream events, the resulting output of signaling pathways could be transcriptional activation/inhibition, cellular events such as cell activation or migration, or physiological events such as blood clot formation.

The earliest pathway visualizations were just hand-drawn diagrams. They showed components of the pathways and reactions linking them together, but with no interactive functions at all. With the emergence of internet and web-based online databases, a number of research groups implemented online pathway databases that display pathways as images with hyperlinks leading to detailed information for the pathway components. More recently, fully interactive visualization of pathways was introduced with some pathway analysis tools, in which pathway components and edges are linked to detailed information, users can drag and move components, and additional data (such as expression profiling data) can be displayed along with the pathways.

For metabolic pathways, the standard visualization layout is to display the metabolites as nodes and the chemical reactions as the edges (using arrows to show the direction of the reactions) of the networks. The enzymes that catalyze the reactions can be treated as the attributes of the edges (Figure 6.4a). Compared to metabolic pathways, the visualization of signaling pathways is more complicated. Even though the basis of signaling pathways is still biochemical reactions, it is the activation, relay or blockade of signals through the pathways that is the most important message to be delivered. To simplify the layout of the pathways, most tools choose to skip the biochemical details and only show the signal transduction (Figure 6.4b). In this simplified layout, network nodes represent the gene products involved in the pathways, ignoring the reactions occurring on the protein. Edges simply indicate the passage of the signal (activation of next component, normally shown as arrows) or the block of the signal (inhibition of a gene product's activity, shown as an arrow with a flat head). There are various biochemical bases for the activation and inhibition events, such as receptor binding, phosphorylation, protease cleavages, translocation in cellular compartments, etc. However, there is currently no standard about how to encode this information in the pathway visualization, and different tools often choose different approaches or choose to ignore all the details. For example, some tools put +p or –p next to the edges to indicate phosphorylation or dephosphorylation, while other tools choose to display nodes in specific shapes to indicate the type of gene families, and users are able to guess the underlying biochemical reactions. In addition to nodes and edges, most tools also use a few other graphic notations to make pathways more understandable and more informative, such as boundary drawings to show subcellular locations, text boxes for additional pathways or environment factors, etc.

Lack of a standard in pathway visualization, especially for signaling pathways, has been a problem for both users and pathways data producers. Different databases using different notations makes it more difficult to understand pathway data and more difficult for data exchanges between different tools. As a result, a consortium involving most of the pathway data providers was formed and released a suggested standard for visual representation of pathways (Le Novère *et al.*, 2009). This standard, the Systems Biology Graphical Notation (SBGN), included three specifications describing three types of diagrams that can be used to display different aspect of the pathways. The first diagram, the SBGN process diagram, focuses on the biochemical and biophysical processes forming the pathways. This diagram is equivalent to the traditional metabolic pathways, with each node assigned to one state of a pathway component and an arrow linking two nodes indicates the transition between the two nodes. A gene product that catalyzes, activates, or inhibits a process would be displayed as a unique edge that goes from the gene product node to the process edge. The second diagram, the entity relationship diagram, focuses on the influences that one gene product or compound posed on another player. In this diagram, an arrow or other edges linking two nodes represents the relationship between these two nodes, such as activation and inhibition, while an edge linking one node to another edge represents the relationship between the node and the interaction represented by the other edge. With the detailed process information removed, this diagram is simpler and less

(a)

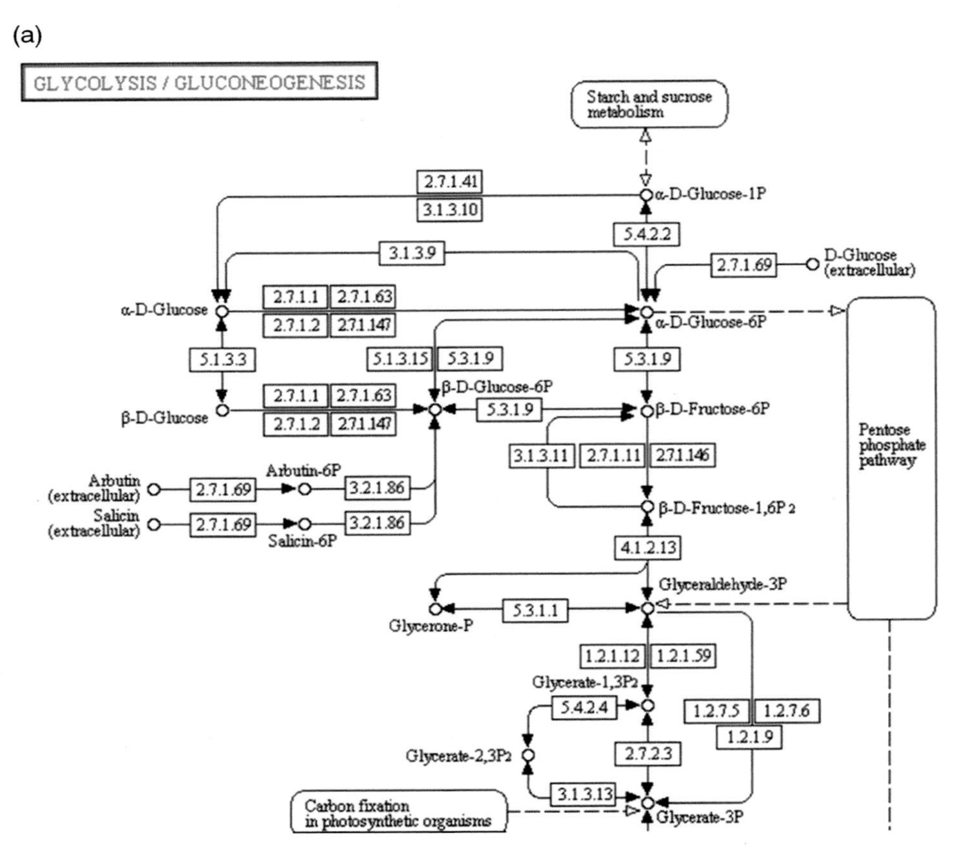

Figure 6.4 (a) Metabolic pathway and (b) signaling pathway shown from the KEGG pathway database. In metabolic pathway visualization, arrows indicate the transformation of metabolites and the boxes next to the arrows are gene products that catalyze the reactions. In signaling pathways, arrows indicate activation of downstream components and flat head edges indicate inhibition or deactivation of downstream components. In this diagram, details of phosphorylation events are added: + for phosphorylation and – for dephosphorylation, y for tyrosine and s for serine.

clogged than the process diagram. The third diagram, the SBGN activity flow diagram, is derived from the traditional signaling networks, just showing the flow of the biological signals. In this diagram, only edges linking gene products are kept to represent the passage or blockade of the signal, making it the most abstract form of the three. Within each diagram, the SBGN uses a defined set of standard notations, such as interaction types encoded in the shape of edges. However, node shape, color, and additional graphic elements such as frames for cellular compartments are not defined in the standard and application developers can use these elements for specific uses (Figure 6.5).

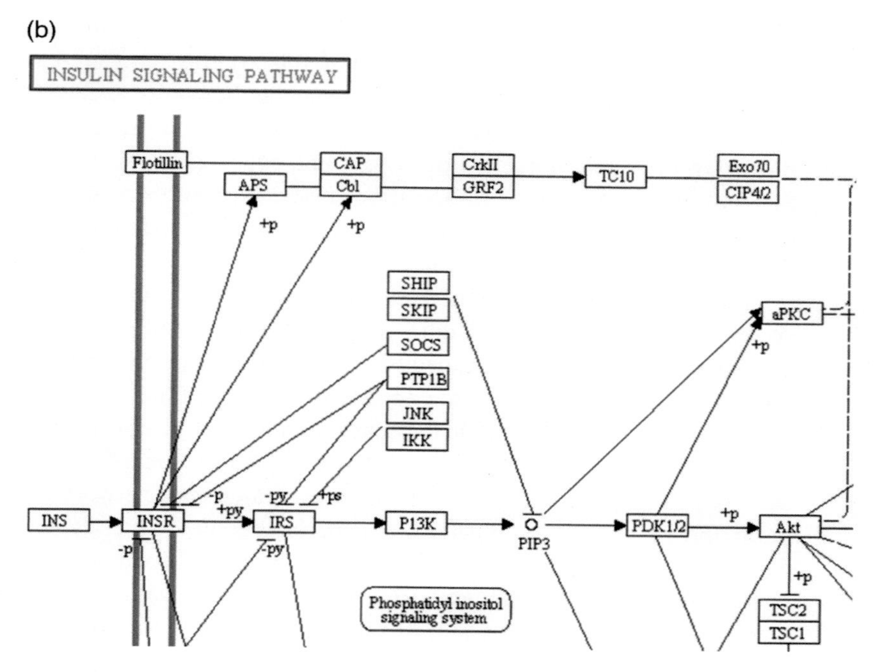

Figure 6.4 (*cont.*)

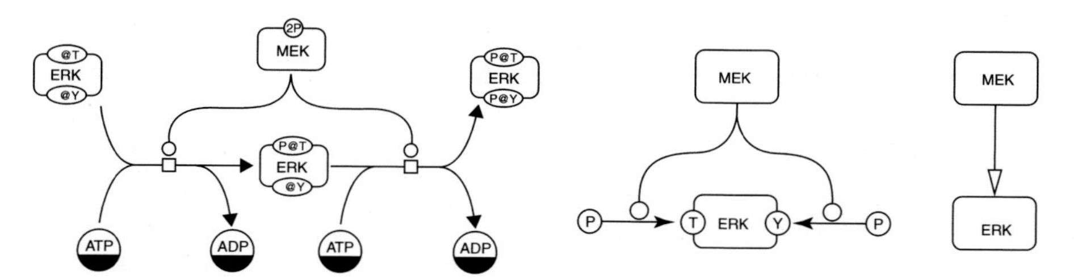

Figure 6.5 SBGN diagrams for one pathway. The process diagram (left) shows the details of the biochemical reactions; the entity relationship diagram (middle) shows the relationship between individual components; and the activity flow diagram (right) shows the flow of the signal in this pathway.

Visualizing molecular networks

There are two components in molecular networks: nodes, and edges. The nodes in molecular networks are gene products or other molecules, such as metabolites or drug-like chemical compounds. The edges are interaction data between the node molecules. These could be data generated from experiments, such as protein–protein interaction data from genome-wide binding experiments, or they could be generated from database mining activities, like genomic sequence analysis or literature mining. One example of this are the literature-based gene–gene interaction data marketed by companies such as Ingenuity, GeneGo, and Elsevier which contain interaction types such as phosphorylation, transcription regulation, protein–protein binding, etc. Both nodes and edges can have attributes. For nodes, they could be gene family information, tissue expression pattern, etc. For edges, they are usually more details about the interactions, which could be reference for the interactions, interaction values, etc.

Visualization has important roles in network data analysis. Through visual representation of the network data, users can easily identify point of interests, such as nodes that are densely connected to other nodes (the so-called hubs) or subregions of the network that are concentrated with certain attributes, etc. Network visualization tools have been in existence for a long time. For most biologists, however, network data analysis started with the Ingenuity, GeneGo, and Elsevier tools that are provided for the analysis of literature-based gene network data. Since then, more and more functional genomics experiments have generated a large body of data sets and a lot of software tools were developed for network data analysis. Among them, a freely available tool, Cytoscape, has gained wide popularity by providing a lot of powerful features, many of which come from user-contributed add-ins.

Unlike pathways, which are mostly developed manually and the locations of nodes on the visualization are fixed for better representation of the biological processes, the layout of molecular network data is not coded in the original network data and is generated by the network analysis software. Different algorithms would generate different layouts, fitting different analysis purposes. For example, Cytoscape provides a large number of layout options. Some of the common ones include: a grid layout, in which nodes are displayed in a square grid; a force-directed layout and a spring-embedded layout, both treating the nodes as physical objects and edges as some forces to calculate the layout of the network which is in a state with minimum sum forces; an attribute circle layout and a group attribute circle layout, which plot nodes with the same attributes in circles. Among them, force-directed algorithms tend to generate a layout with more balanced node distribution, grid and manual layouts are better for simple networks, while attribute circle layouts are particularly useful for identifying patterns related to certain node attributes. In addition to these layouts for standard network data, biological data have some attributes that are unique and it is possible to create biology-specific layouts. For example, many gene products have subcellular location information and many tools implemented a network layout which can display networks in layers based on subcellular information, such as extracellular, cell membrane, cytoplasmic, nuclear, etc. (Figure 6.6).

In addition to network layout, most tools allow users to encode node and edge attributes into visual properties, such as node shape, color, size, transparency, etc. Cytoscape has very rich features in these options, which makes it very convenient for network data analysis. For example, setting the node size based on the edge

(a)

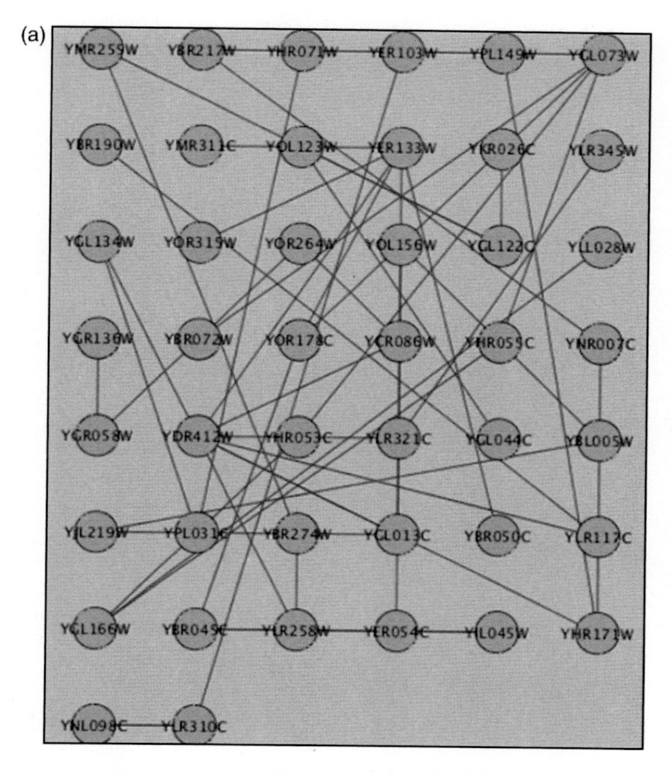

(b)

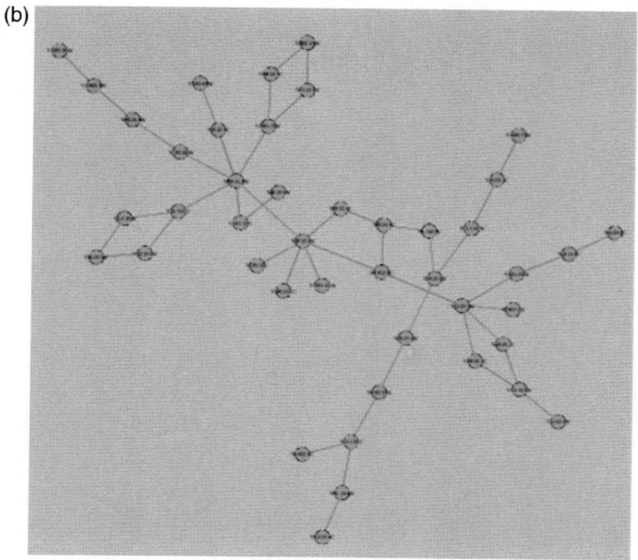

Figure 6.6 Several network layouts: (a) grid layout; (b) force-directed layout; (c) attribute circle layout, taken from Cytoscape User Manual (www.cytoscape.org); and (d) subcellular localization layout, created using Elsevier's Pathway Studio.

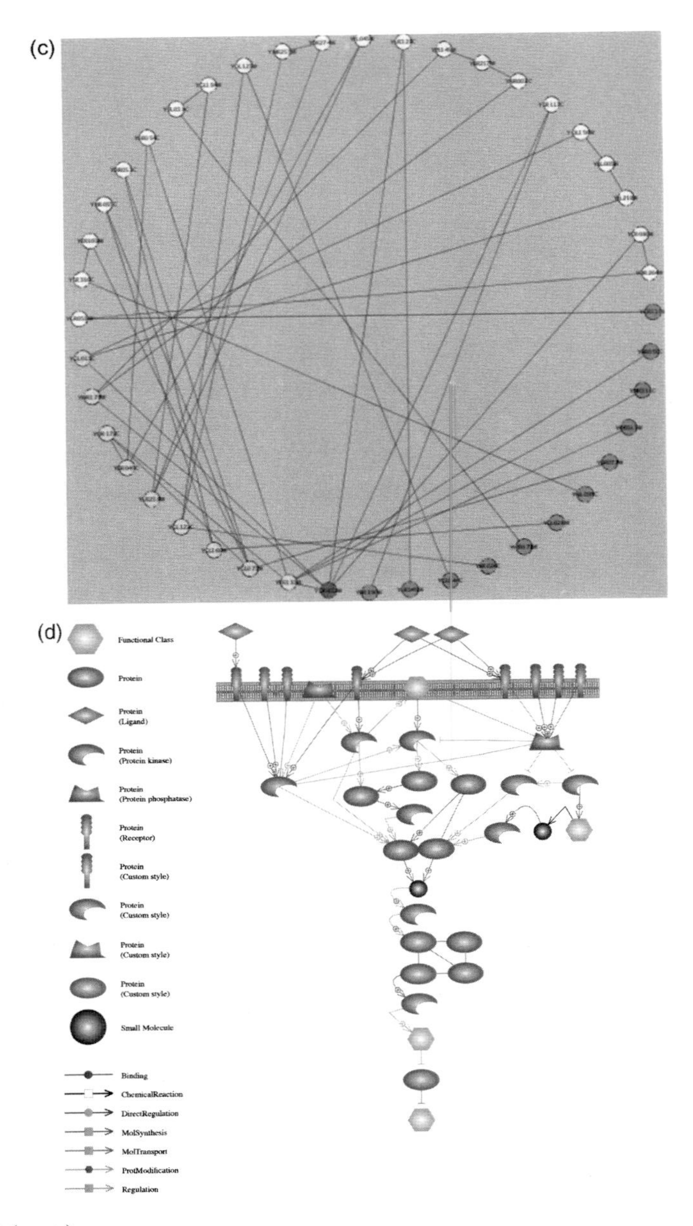

Figure 6.6 *(cont.)*

numbers makes it very easy to identify hubs. Node color based on expression data is almost a universal representation method for visualizing gene expression data in network settings. Some convenient additional visual encodings are: node shapes for gene family information; edge line styles for interaction type; edge thickness or opacity for interaction strength, etc.

Functional genomics and other numerical data

Numerical data have long been part of biological study results, but it was only after the birth of microarray-based functional genomics technologies that visualization of the data using software tools became routine tasks for biologists. This is mainly because the size of the data sets generated by these experiments is simply too large to be viewable without the assistance of visualization tools. Microarrays for gene expression profiling studies typically measure the expression of thousands to tens of thousands of genes at a time. Even the genes changed significantly in an experiment are usually in the range of hundreds or more, which is still beyond the numbers viewable without assistance of visualization technologies. Genotyping arrays, on the other hand, typically monitor hundreds of thousands to more than one million SNPs on each array, generating even larger data sets. In the early days of microarray studies, there were not many visualization features from the data analysis software tools provided by the array manufacturers. As a result, general visualization tools, such as Spotfire, gained huge popularity in research communities. Subsequently many microarray data analysis tools started providing more and more visualization features. Tight integration of the visualization functions to the underlying data analysis process made these tools quite attractive.

Microarrays for gene expression profiling can be classified into two categories: single-channel arrays with one hybridization sample, and dual-channel arrays that take two samples labeled with different dyes in the hybridization. The output data of single-channel arrays are hybridization intensity values that reflect the expression level of the gene. Dual-channel arrays usually use one of the channels as internal control and output a ratio between the query sample and the control sample for each of the genes on the array. With multiple samples used for each experiment, the raw data for one experiment can be organized as a tabular data set, with each row representing a gene and each column representing a sample (either signal intensity for single-channel arrays or a ratio for dual-channel arrays). This type of data set is also called high-dimensional data, in which each sample is considered a dimension. In addition to these values, columns of gene annotation data may be appended to the data set, which makes the dimensions even higher. Genotyping data are typically converted to the association P value, which has only one dimension. However, with gene annotation data, they become high-dimensional as well.

Many visualization methods have been developed for multi-dimensional numeric data. Some are better suited for visualizations of whole data sets, while others are more appropriate for visualizing individual data points. Next, we will discuss some of the most widely used visualization methods and their usages in functional genomics data analysis.

Visualizations for whole data sets

The main reason for visualizing the whole data set is quality checking. Errors and problems can happen in many steps of a microarray experiment that cause bad data quality and outliers. Some of the problems can be identified by data analysis tools, and some are easier to identify through visualization. For example, a plot of an Affymetrix chip probe intensity on the chip physical layout is a powerful technique for identifying problems caused by scratches on the chip or uneven hybridization.

For treatment groups with more than two samples, pair-wise plots can be displayed in the same graph as small graphs (the so-called small multiples). With each small graph acting as the control for the others, a sample with overall expression pattern different from others can be easily identified as the dotted area further away from the diagonal than others (Figure 6.7).

Visualizing individual genes

One of the main purposes of gene expression profiling experiments is to identify genes significantly changed between different sample groups. Visualization of one gene across all samples, usually grouped by each treatment group, is an important part of end results. It is done normally by using the x-axis for treatment groups and showing expression values on the y-axis. This is the simplest form of profile plot. There are many ways of displaying the values. They could be displayed as dots, bars, or combined forms such as box plots. If the treatment groups are time series, it is quite often shown as a line graph, in which a line connects the mean of each time point. Sometimes, the experiment is designed as a multiple factor experiment, and one of the factors can be chosen as the x-axis and other factors may be encoded in other visual properties, such as color of data points.

The implementation of the above concept varies among different tools. The microarray analysis tool array studio implemented all these features in an easy-to-use view called the variable view (Figure 6.8). After the microarray data and experiment design table are entered into the tool, users can launch a variable view to show the values of any genes with a few clicks.

Visualizations of gene clusters and expression patterns

Gene clusters in expression profiling experiments are groups of genes that show a similar expression pattern across all samples in a given experiment, which indicates they are possibly regulated by the same transcription factor and/or the same signaling pathway. These can be identified by clustering analysis, such as hierarchical clustering or K-means clustering, and they are typically visualized with heat maps and profile plots.

Heat maps show the expression of values of n genes across m samples in a grid plot with n rows and m columns. Each gene in one sample is shown as a square in the plot, with the color of the square representing the expression value. Typically, the expression value is converted to the log ratio between the treated group and the control group, with zero meaning no change, positive values

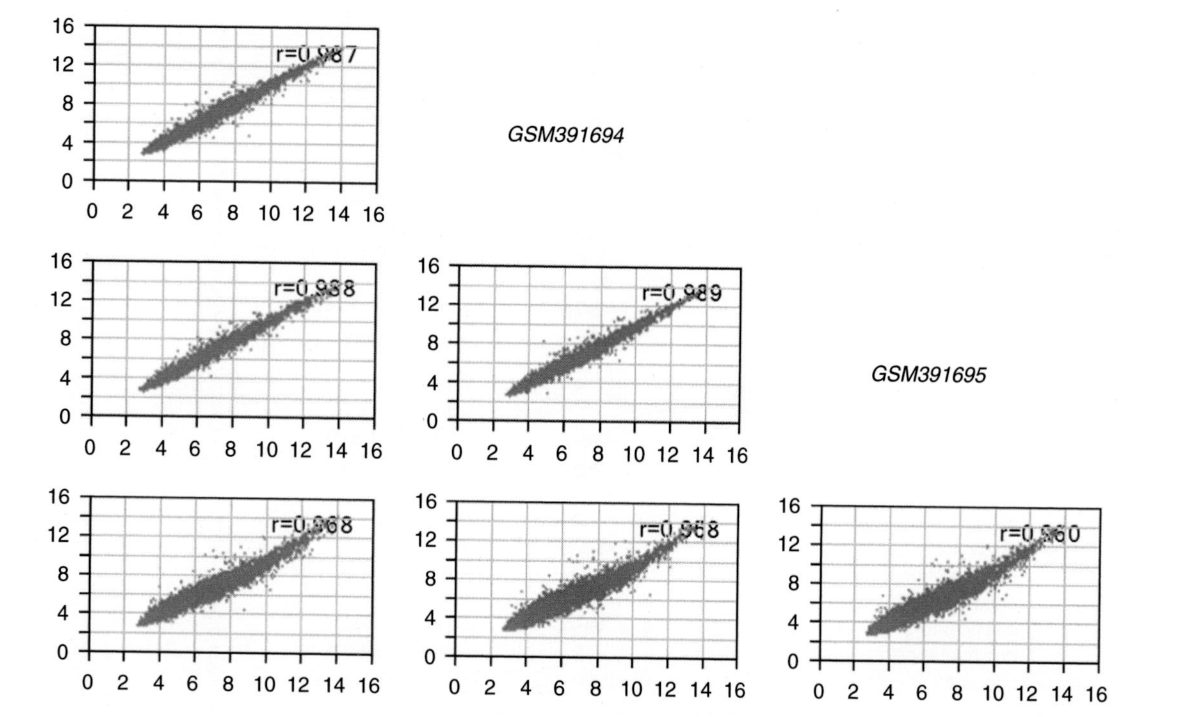

Figure 6.7 Visualization of whole data sets. Small multiple scatter plots showing pairwise comparison of several samples from one experiment.

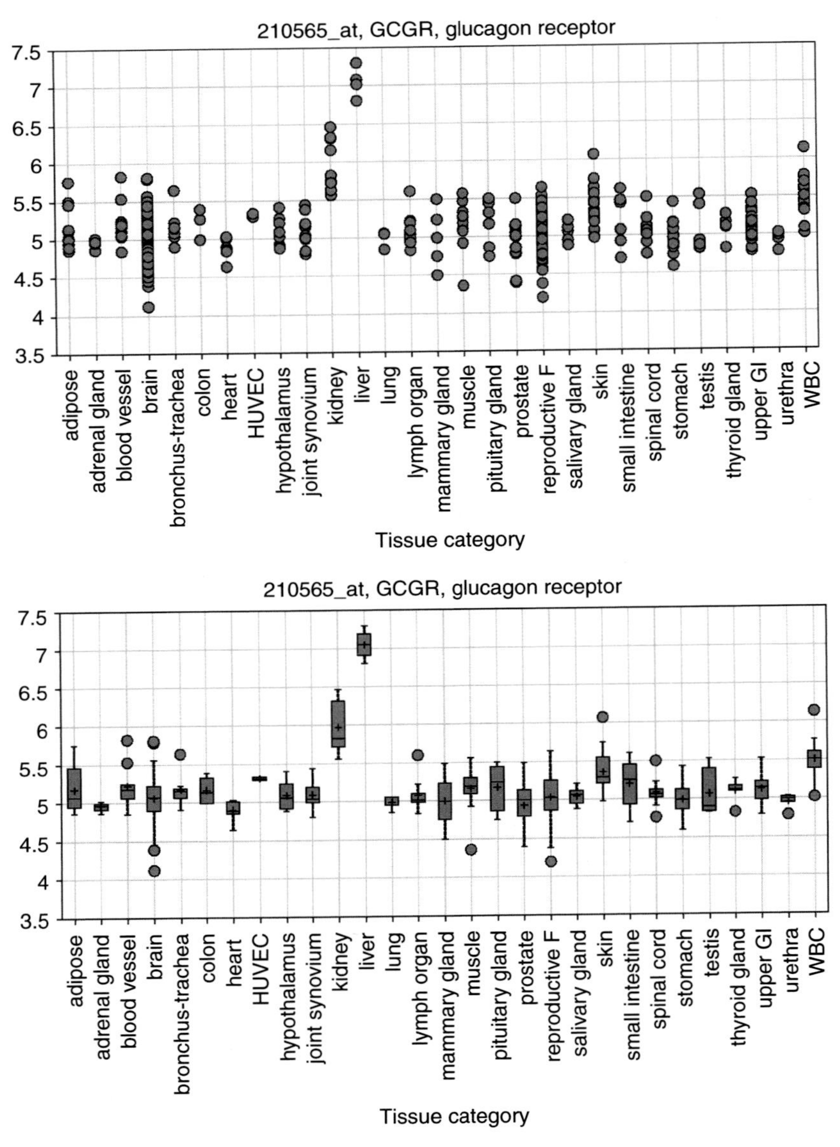

Figure 6.8 Two variable views showing expression pattern of one gene across different tissues.

meaning upregulation shown as red, and negative values meaning downregulation shown as green. Heat maps can be used to show genes and samples in the original order, but more likely show the clustering result, in which similar genes and similar treatment groups are rearranged next to each other. With the color coding, users can easily identify areas of similar colors which indicate genes with similar expression patterns.

In profile plot graphs, the y-axis is used for expression value and all genes have to be displayed in the same area. With this limitation, it is best to display different clusters in different plots. This is usually achieved by the K-means clustering algorithm, which clusters genes into K groups based on expression pattern similarity. The expression pattern of each group is usually shown as a line graph and all K line graphs together represent the clustering results.

Visualizing data in biological context

The ultimate goal of functional genomics experiments is to map the impacted genes or SNPs to biological contexts such as pathways, functions, or chromosome locations. Multiple genes in one pathway changing in the same direction, multiple SNPs in the same chromosomal region showing significant genetic association, could drastically increase the confidence of the results. Visualizing the data under these biological contexts is, therefore, a very important part of functional genomics data visualization.

Most of the pathway and network visualization tools provide features to display expression values in the diagrams. Usually, expression values are shown in the color of the nodes or as colored visual elements next to the nodes, with red for upregulated and green for downregulated changes (Figure 6.9). Some tools even implement more detailed visualization to display expression pattern. One fine example is the tool VANTED (Junker *et al.*, 2006), which can show expression values in many forms or visual styles, creating a pathway diagram with very elegant small multiples (Figure 6.10a).

For SNP chip-based genome-wide association data and copy number variation data, plots with chromosomal location on the x-axis is the standard visualization layout. With this layout, users can easily see the regions that are significantly associated to some phenotypes and regions that are lost or amplified (Figure 6.10b).

Concluding remarks

In this chapter, we have shown visualization methods and software tools for three basic data types in genomics and functional genomics. These were chosen largely because they compose the majority of the large-scale data sets generated so far. As the genomics research matures and evolves into systems biology, data integration of more data types and larger scales will require the development of new algorithms, and more user-friendly tools. Meanwhile, with the growth of other data types, such as protein structure data, bio-image data, text data, we will likely see new visualization technologies for those data types in the near future.

(a)

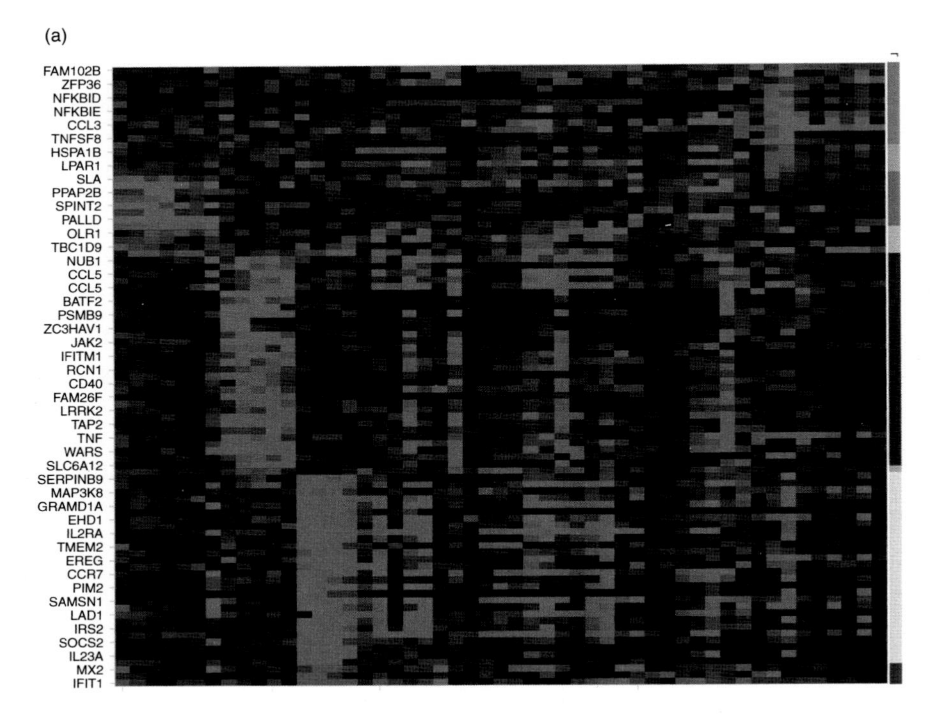

(b)

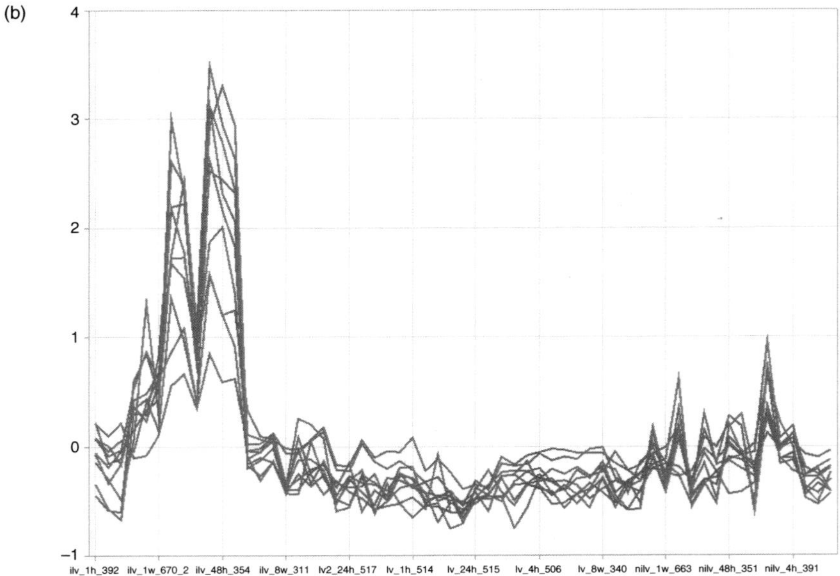

Figure 6.9 Heat map (a) showing a group of genes with expression pattern correlated to cancer metastasis status and profile map (b) showing clusters of genes with circadian expression pattern. A black and white version of this figure will appear in some formats. For the color version, please refer to the color plate section.

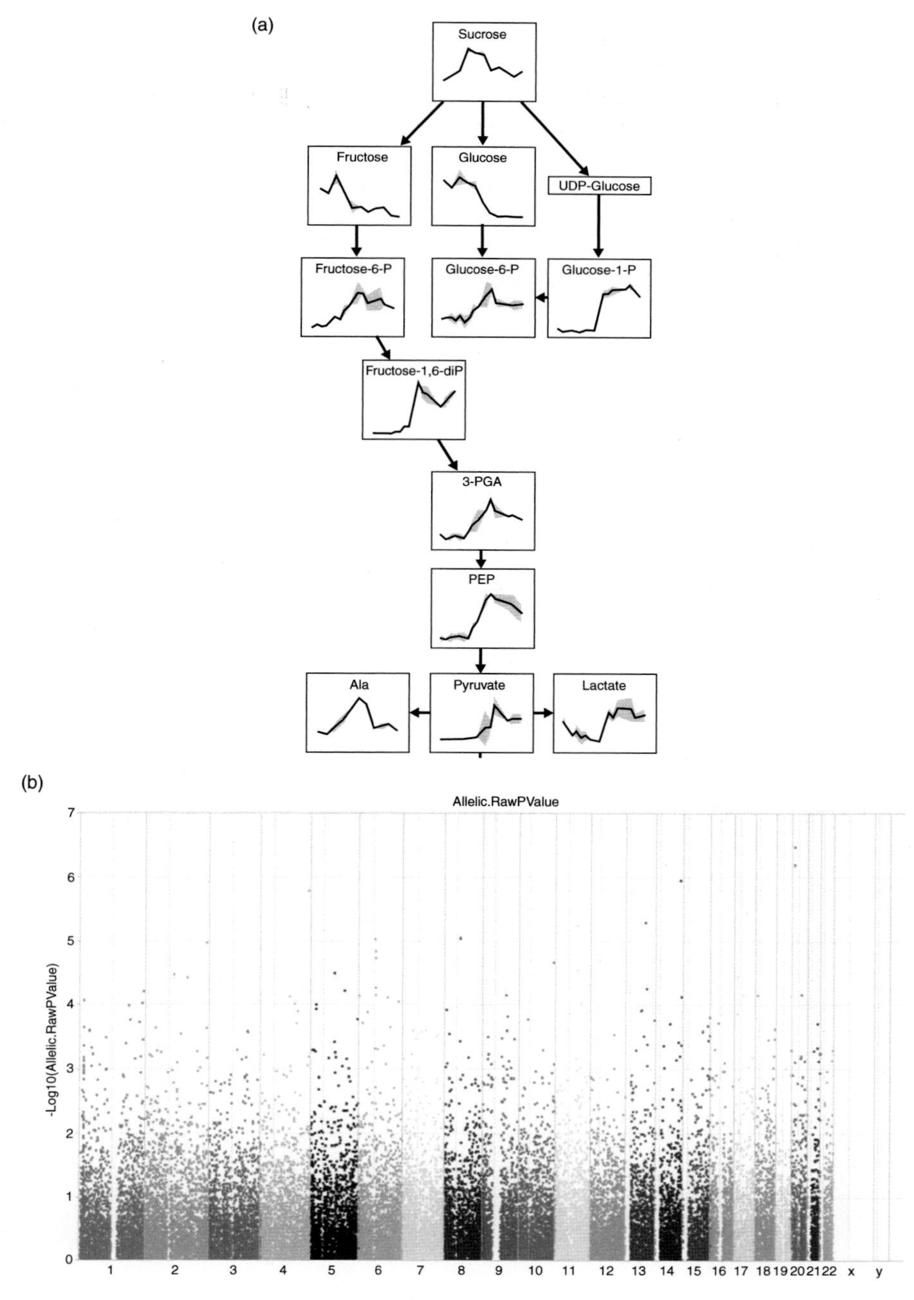

Figure 6.10 Visualization of numerical data under biological contexts: changes of metabolites under a metabolic pathway layout (a) and *P*-value of genetic association data on chromosome layout (b).

References

Berriman, M. and Rutherford, K. Viewing and annotating sequence data with Artemis. *Briefings in Bioinformatics*. 2003;4(2):124–132.

Frazer, K. A., Pachter, L., Poliakov, A., Rubin, E. M. and Dubchak, I. VISTA: Computational tools for comparative genomics. *Nucleic Acids Research*. 2004;32: W273–W279.

Junker, B. H., Klukas, C. and Schreiber, F. VANTED: A system for advanced data analysis and visualization in the context of biological networks. *BMC Bioinformatics*. 2006;7:109–121.

Le Novère, N. L., Hucka, M., Mi, H., *et al.* The systems biology graphical notation. *Nature Biotechnology*, 2009;27(8):735–741.

Lewis, S. E., Searle, S. M., Harris, N., *et al.* Apollo: A sequence annotation editor. *Genome Biology*. 2002;3(12):research0082.1-0082.14.

Schwartz, S., Zhang, Z., Frazer, K. A., *et al.* PipMaker – A web server for aligning two genomic DNA sequences. *Genome Research*. 2000;10:577–586.

7 Information visualization – important IT considerations

Telmo Silva

Innovative approaches in drug discovery must be accompanied by innovative approaches in technology supporting those approaches, especially in the area of information visualization.

This chapter outlines the five fundamental components of computer-aided information visualization (CAIV). The components are: hardware components, interaction components, data visualization, application design, and user adaptation. These are not arbitrarily selected components, but rather an attempt to structure and assist with the effort in effectively implementing information visualization in an organization or department.

Additionally, this chapter will focus on the current and future state of each of the fundamental components that have been identified rather than its background. However, it is my hope that the topics will raise enough interest in the reader to further investigate and research the foundation of each component in its historical context to better understand their future.

Finally, this is an attempt to provide a novel and useful structure to assist a scientific or business organization with defining their implementation focus and effort, hoping that the readers will leave with a clear path on putting in place a proper approach to data mining and visualization in their academic or professional field, specifically in bringing more products to their final stages of development.

It is impossible to cover in detail in only one chapter the topics outlined above, but this chapter can be seen as a small introduction and reference guide to many other sources of information for the reader, so that they can start to understand and discover the potential and promise of CAIV for their organization and projects.

From data visualization to computer-aided information visualization

A distinction between data visualization as one of the components of CAIV is required to begin this chapter. Data visualization is in itself the core of CAIV where poor data representation design will undermine the whole analysis process. As Edward Tufte stated, "Graphical excellence consists of complex ideas

Bioinformatics and Computational Biology in Drug Discovery and Development, ed. W.T. Loging. Published by Cambridge University Press.

communicated with clarity, precision, and efficiency" (Tufte, 2007). He adds, "Graphical excellence is that which gives to the viewer the greatest number of ideas in the shortest time with the least ink in the smallest space."

It is here that CAIV enters and adds the remaining four components to facilitate the viewer to navigate, interact and react to the data presented. Due to the higher complexity, volume and requirements of today's analytical needs, these new components are as important as the data visualization piece itself. Without them, the best-designed graphical representation will not be seen, or worse, will contain incorrect, incomplete, or misleading data, or in Tufte's words, "lies."

As stated previously, the five components of CAIV are:

1. Interaction components
2. Hardware
3. Application design
4. Data visualization
5. User adaptation

Göbel, Müller and Urban have created a structure of visualization topics in a similar fashion as it relates to scientific and three-dimensional (3D) data representation (Göbel *et al.*, 1995).

A new framework

Creating a new framework by which we implement information analysis is critical, because previous assumptions and methods do not realize the full potential of information visualization. As mentioned in the introduction, restructuring the implementation of analysis systems in the five elements will lead to an overall improvement in any area that uses CAIV.

Why five elements? The reason for the above five components is based on the objective of information visualization. Information visualization is an important method in providing a person the ability to easily and quickly navigate through large amounts of data, recognizing data patterns (links, trends, cause and effect, transformations, seasonality, etc.). By examining how this method can be implemented to serve scientists performing an investigation, we have mapped out five components.

CAIV as it relates to science has been a topic of interest and investigation since the early 1950s. It was not until the early 1990s, however, that the scientific community initiated events such as the Eurographics Association visualization conferences and the IEEE Visualization Conferences taking data visualization to the computerized age (Post *et al.*, 2003). CAIV also became of greater interest to the non-scientific community, but real-world examples have not been available to the general public until early 2000.

Perhaps the best public example to date of CAIV, exhibiting all five elements, is the 2008 US Presidential Election broadcast coverage. What CNN called the "Magic Wall" is a simple but complete illustration of the five elements of CAIV where John King used a multi-touch panel to explain to the viewers a series of election data using a variety of maps, charts and diagrams.

If we break this example down into the five elements, the large screen and computer systems behind it represent the hardware element; the multi-touch capability of the screen is the interaction element; the charts, maps and diagrams are the data visualization elements; the fluid animation and navigation are the application element; and finally, Mr. King himself is the user element.

The framework enables John King to convey and explore data alongside millions of viewers in a simple and highly effective way. If this was not used in a televised event, additional findings and interesting relationships could have been identified using this CAIV, instead of merely showing the carefully scripted analysis flow.

Although interesting, the election coverage is simple in comparison to drug research or business analysis, and was limited to collecting well-known data, i.e., counting votes for known candidates from a known geography with known expected result possibilities, but it serves as an example of the CAIV framework.

It is important to note the difference between information visualization and data warehouse building or data mining.

In building data warehouses or data marts (as the names imply), one pre-selects the data before storing them in an accessible location and displaying them using data visualization components. This type of data filtering prior to visualization means that unknown relationships and following new paths of investigation and analysis are not possible.

Typical data warehouses promise the ability to analyze information and make decisions based on displayed information. However, investigating the processes of data warehouse implementation in organizations, the first step in any successful implementation is to identify the user's questions and analysis needs. In fact, looking at the advice of leading experts in the field and literature, the first phase of implementing data analysis and data mining tools is actually to select and transform the data set (Soukup and Davidson, 2007). It is a contradiction that in the search for new data relationships, first the relationships have to be built from a data set limited by design.

The above framework is the one that most scientific and business communities base their analysis on, in defined and structured silos of information where data are prepared, transformed, cleansed, and linked prior to the first visualization of the same data set.

Hence, we arrive at the need to implement true information analysis systems, complete with the possibility to build relationships between data elements on the fly, relate cause and effect functions, visualize complex nodal trees in 2D or 3D, and navigate through the data to better understand it at a faster speed with more data connection points than ever before.

Information visualization is truly the next generation of what we today call business intelligence or data warehouse building. The scientific community has already been doing the process of information visualization with the aid of disconnected software applications and disparate cleansed data sources. However, the time and effort for such an analysis is perhaps the biggest challenge that CAIV needs to solve.

Hardware

The ever-increasing amounts of data demand that newer and better hardware be utilized. Of course, there is hardware such as Silicon Graphics' Stokes supercomputer, one of the top 200 fastest machines existing today, powered by 640 Intel® Xeon® processors (totaling 2560 cores) and features 5.1 terabytes (TB) of memory and 84 TB Panasas ActiveStor parallel file storage solution.

While that may impress some, considering the power that the average consumer has, this is no longer impressive. In fact, the above is no longer constrained by money; it comes down to having enough space, electricity, and cooling equipment to maintain the 6 tons of hardware that makes up this system.

A typical household with two computers (perhaps a desktop and a laptop) plus attached peripherals (backup drives, iPods, memory sticks, TiVo, etc.) already has two 2 processor core equivalents, and approximately 1.5 TB of storage space and 10 GB of memory. A small company with fewer than 10 employees will typically have between 10 and 15 TB of shared storage space and close to 160 processor core equivalents.

So what does this mean? It means that the supercomputer still has a place in focused scientific communities, but grid computing or distributed computing can be a much more powerful, accessible and affordable way of adding power to your research facility.

Consider the investment in computers that a typical midsized research, biotech or pharmaceutical makes in equipping all their employees with two- and four-core processor computers, which could be around 5000 core equivalents. Harness that processing power for all the computers which are idle between 6 p.m. and 6 a.m., and you have an idea of the potential that distributed computing can provide.

The number is astonishing, and the potential for additional return on investment in computing equipment could triple if departments that require computing power to process data can now use the dormant processors (and their local storage) for temporary distributed calculation and analysis.

Consider the alternative, purchasing a supercomputer for a price tag of US$1 million or more, where people (not computers) are the ones taking turns to use it to run their research calculations. Limiting the use of computing power based on the availability of processing power is reminiscent of the early days of supercomputers.

Nonetheless, SGI's Stokes supercomputer gives us an idea of what research departments should aim for with respect to hardware requirements: a system with multi-processors, multi-cores, a large memory capacity and large amounts of storage. This can be achieved at a mid level using Microsoft Windows HPC Server 2008, which by incrementally adding hardware to this 64-bit operating system, Cray-like processing power can be achieved over time. With the Windows HPC

2008, thousands of processing cores can be managed centrally, maintaining system availability at a peak, and it allows for server and data resources (Microsoft Windows Server and Microsoft SQL Server) to be utilized in cluster pools. The path is set to utilize any processing core, including future workstation operating systems such as Windows 7, which relies on much the same architecture as Windows 2008, to be used as a possible operating cluster.

Bringing typical workstations into a parallel computing environment brings additional advantages. Typically, servers are not well-equipped when it comes to graphic/video hardware. High-end range professional workstation systems can offer perhaps even better visualization capabilities as well as power, because they have superior graphics capabilities.

Bridging data processing with visual projection is one way to improve performance of data processing and rendering algorithms, called GPU computing. The model for GPU computing is to use a CPU and GPU together in a heterogeneous computing model. The sequential part of the application runs on the CPU and the computationally intensive part runs on the GPU. From the users' perspective, the application runs faster, and they do not care that it is because the application can utilize the GPU to boost performance.

The application developer has to modify their application to take advantage of the GPU, and offload processor-intensive calculations. The rest of the application would remain using the CPU. Mapping a function to the GPU involves rewriting the function to manage the parallel processing in moving data to and from the GPU and CPU.

GPU computing is enabled by the massively parallel architecture of NVIDIA's GPUs called the CUDA architecture. The CUDA architecture consists of hundreds of processor cores that operate together to crunch through the data set in the application. Again, the implications are that the other areas of CAIV need to be adapted to work with this new technology; i.e., applications need to be redesigned to be able to distribute the work among the processors (both CPU and GPU). Once such example is NVIDIA's CUDA architecture.

> Although GPUs have been designed primarily for efficient execution of 3D rendering applications, demand for ever greater programmability by graphics programmers has led GPUs to become general-purpose architectures, with fully featured instruction sets and rich memory hierarchies. NVIDIA's CUDA presents to the programmer a fairly generic abstraction of a many-core architecture supporting fine-grained parallelism. CUDA and the GPU therefore provide massive, general purpose parallel computation resources with the potential for dramatic speedups. (Boyer *et al.*, 2009)

Already some results are available in scientific research.

> We have shown that leukocyte detection and tracking can benefit greatly from using a CUDA-capable GPU. The algorithms used in the detection and tracking stages, stencil computations and iterative solvers, are also used in a wide range of other application domains, which can all benefit from the optimizations we have discussed. Overall, the best CUDA implementation provides speedups of 58.5× and 211.3× on the detection and tracking stages, respectively, over the original MATLAB implementation and 9.4× and 27.5× over the best

OpenMP implementation. While the MATLAB implementation takes more than four and a half hours to process one minute of video, the CUDA implementation can process that same video in less than one and a half minutes. (Boyer *et al.*, 2009)

The CUDA GPU architecture and the corresponding CUDA parallel computing model are now widely deployed with hundreds of applications and nearly a thousand published research papers.

Corporations involved in research and development, especially in the pharmaceutical domain, must go back to the roots of scientific discovery and unleash another era of computing growth, similar to the 1970s and 1980s and that eventually allowed for the creation of sophisticated and affordable computers; through the research done in Palo Alto by Xerox, and the development of ARPANET by the scientific community which led to what we know today as the Internet.

Today's scientific community needs the support of their information technology partners to assist in increasing their budget for memory or bandwidth usage, by altering their policies on hardware standardization to allow for the latest innovations, and moving forward based on the benefits of implementing new solutions rather than hanging back based on fear of security issues, to properly utilize the amazing power of higher-end powerful workstations capable of using idle network resources with large data transfer capabilities. Information technology departments can focus on implementing standardized solutions to maximize the appearance of safety rather than trying to implement performance-driven standards safely.

Input/output devices

Looking at the timeline of the wide use of computer–human interaction devices, it is clear that there is a need for supporting or alternative ways of interacting with computing systems, specifically in navigating in 2D and 3D data models. The keyboard, first put to use in the post-World War II era, has remained, for close to 70 years, the preferred method of computer input device. The mouse has likewise reigned for close to 30 years without dispute and will most likely continue for some decades to come.

However, changes in the other information visualization areas will ensure that alternatives will soon be designed and implemented. In effect, this has already started in the areas of personal computing and in the electronic game industry. Not only is there a noticeable delay in bringing new interfaces to the market, but more importantly there is a lack of acceptance by older generations to use devices that may force them to stand up and touch screens, perform gestures resembling dance moves, or voice commands to a computer using keywords.

Voice recognition is another input method which has been around since the early 1990s and yet has not gained wide acceptance. As interesting as it is, voice recognition is widely used in automated phone queues, phone surveys and managing your voice mail box, and yet few people actually use voice recognition to interact with their computer. Most people are still much more comfortable with a manual approach. Yet Dragon Naturally Speaking has been very successfull in bringing voice recognition to the market, and very few people know that since Windows 98, voice recognition and voice output is a part of the Windows operating system standard set of functionality. Voice recognition is still severely under-used, not because of its capability but rather due to the hardware placement. Microphones require close placement to the user's output device, the mouth, or sensitive enough to pick up the commands of the user without the surrounding noise of other people's voices. Additionally, it is difficult for current voice recognition systems to differentiate between a command or a simple thought said out loud, or a conversation when someone enters the room.

The Nintendo Wii remote control was developed as an input device using standard and simple components which, when adapted and restructured, are able to navigate through complex 3D models. More interesting adaptations of combined projection and infrared devices are seen through the many projects by Johnny Chung Lee and the amazing ingenuity associated with each one of them (http://johnnylee.net).

The growth of single and multi-touch panels has grown in popularity, mainly due to Apple's iPod/iPhone releases. Apple pioneered the PDA with the Newton which has touch input from a stylus and recognized handwriting. The learning curve for this device was either too steep or its lack of input feedback did not appeal to the average consumer. The Palm PDA devices then dominated the market, until Apple returned with the iPod and iPhone. The industry has now moved away from using a stylus to having the user interact with the screen through touch; this was a logical move, but the industry is divided on how to best adopt touch interfaces with human interaction. Part of the success of touch technology is the fact that the user is now interacting with the screen location of objects; this is also a problem. If the visual output area is small, touching individual items becomes difficult; if too large, then it becomes impractical to walk or reach the visual object to interact with it. Writing also becomes difficult as the projection surfaces are typically not set at the correct angle for the user. When writing on a whiteboard, human beings are faced with an unnatural writing and interacting position. The arms parallel to the body are forced into a curvature and additionally unsupported, which does not allow for precise movement and control. Holding the arms perpendicularly to the body and still unsupported also does not allow for the control required to precisely point to small areas or to write. As such, tablet and table devices are a good alternative to large wall vertical displays.

N-trig's digitizers for tablets are currently the only interface devices in the market to support user input from both pen and capacitive touch in a single device, which can facilitate switching to writing mode more easily, but their current product line is limited to 19 inch/48 cm screen devices.

Perhaps the largest issue with data visualization is the ability to have the overview of many data sets and be able to navigate and maintain several data sets in view at all times; 20- and 22-inch screens do not provide enough area for multiple data sets to be viewed simultaneously. In fact, video cards containing two or three outputs are becoming more common these days, as we see more dual-screen displays occupying desk space. Interestingly enough, financial traders and hotline/call center support specialists are leading the wave of technology adoption while business and scientific analysts in many pharmaceutical companies are still confined to their single monitor to view their data.

Perhaps it is time to expand conventional thinking into encompassing the use of wall displays (50-inch screens or LDP projection systems) which can better utilize vertical space in the office and also invite teams to contribute and participate together in data analysis. In fact, combined with intelligent projection systems, any flat surface will do.

The interface made famous by the movie "Minority Report" is of course the ultimate goal of many companies such as Atracsys, Perceptive Pixel founded by highly influential industry leader Jeff Han, and Oblong Industries.

Atracsys' portfolio of touch screens is gaining some popularity with marketing departments to showcase their products. Zimmer, an implant manufacturing company, showcased their new spine implants using this technology which provides better visualization of their product.

The interface interacts extremely well with single touch-based manipulation and has the advantage of transparent projection surfaces. This Swiss company continues to be focused on optical tracking with a strong medical research background.

Oblong Industries is the developer of the "g-speak spatial operating environment." One of their founders actually served as an advisor in the movie "Minority Report" and their approach is substantially different than most other simple "multi-touch environments." In fact, their approach since early 1990 is the development of an operating system without a mouse and keyboard (or at least one without a physical set). The creation of a type of sign language that the operating system interprets as commands to replace the mouse is the key objective of g-speak and the demonstration movies on their website show a promising new future in this area.

Perhaps the most well-known of technical researchers in the arena of touch-based graphical interaction devices is Jeff Han. Made famous by his demonstration in TED Ideas Worth Sharing (2006), where his demo of his multi-touch tablet and application was a success, now his company, Perceptive Pixel, builds specialized multi-touch geo-mapping interactive screens and applications for the US Defense Department.

Any time you have this command and control situation or you need to put up lots of data. The military has a big problem right now. They have all this data. They have automated drones flying around with video cameras all providing video feeds. They have satellites flying around all the time taking images of everything and it's hard to sort through them now. So they need tools to help manage this huge, huge influx of photographs and other kinds of reconnaissance.

So having maps that are very easy to use and that you can lay side by side and show how a terrain changes over time. It also really allows high-ranking people to use it to be able to operate it. Rather than a computer expert using it, this is something that a high-ranking commander can use. (*Wired*, Jeff Han Interview, 2007)

Perceptive Pixel provided the "Magic Wall" application and hardware that enabled the coverage of the 2008 elections. Much of the attention on today's interaction is based on the necessity of quicker interaction with the application. Typically, humans are the slower of the two, able to type as fast as their dexterity and hand–eye coordination allows them, or move their arms in gestures or touch specific areas of the screen quickly enough to allow for a smooth interaction.

Some interesting advances in wearable devices are happening as well. The idea that the hardware moves with people may prove less costly and more practical at the expense of having the end-users wear or carry the device itself. A great example of this is the Oculus Rift.

The business and scientific community are now taking a device primarily targeted for gaming into the field of data analysis. A great example of this is in finance, where educational institutions spend upwards of half a million dollars per year building and maintaining financial labs equipped with multi-screen trading room simulators for a limited number of students at a time.

In lieu of that investment, the Oculus Rift provides an interesting alternative where multi-displays are virtually placed inside the stereo display, allowing the student to navigate the trading room at a fraction of the cost.

The main advantages of virtual financial laboratory will be lower cost, reduce the need to regularly upgrade the hardware, provide an environment that can be adjusted to the varying needs of financial exchanges and banks, trading centers, ease and flexibility to manipulate the laboratory environment as per the industry demands. The laboratory will be able to adapt to the different needs of different financial exchanges and banks, trading centers. (Yin, 2014)

The advances in display technology, fixed, mobile or personal, including higher resolutions such as 4K, are well on their way to providing the user with clarity in data visualization.

But how does the application communicate back to the user apart from displaying additional visual data onto the surface? Eventually, despite the size of the screen, the application will run out of space as well unless it can provide information to the users via another medium, such as sound.

It is not only simple beeps and warning noises such as those that occur when starting your laptop. It is hard not to react emotionally when a well-designed sound effect or audio track is executed alongside its image equivalent. Sound creates emotions and can guide people to look in detail at specific areas of the screen. There are also voice emulators which have been improving year after year as showcased by Alex, the Apple Operating System default voice which directs users to look at specific areas of the computer in order to resolve issues. America On Line's "You've got mail" is still heard today in millions of computers. It provides a very quick way for the application to notify the user that there is additional information available, without cluttering or covering already existing information being displayed.

Data visualization

Data visualization can be thought of as the heart of CAIV. Data visualization plays an integral role in the display of information and if it is not properly implemented, the resulting deficiencies will undermine all other components of CAIV.

Yet, since the general availability of computers for business and scientific use, there has been little progress and innovation in this area. As an example, one of the most widely used data analysis tools today and for the last 10 years is Microsoft Excel. Both Excel's datasheet view as well as its chart wizard component changed little during that time span, with the exception of color schemes.

An argument as to the reason for the lack of change could be that charting must follow certain conventions to be useful as an analytical tool, but innovation in this area would provide great benefits to both users and software providers.

With poor data visualization implementation this will just add noise to data rather than simplifying its representation. Experts in the area of data visualization are clear that there are limits to the amount of useful information that can be placed on a screen, and furthermore, there are many tools available to mark up data visualizations, paint the data with additional elements (3D, colors, additional mark-up such as arrows and highlights, etc.), but adding mark-up information should only be used when it adds to the clarity of the data analysis.

As stated by Benjamin Fry, "Because of the accuracy and speed with which the human visual system works, graphic representations make it possible for large amounts of information to be displayed in a small space" (Fry, 1997). This is truer in using computers because one can quickly pan and zoom across many data charts, but beware of Tufte's words:

> Occasionally designers seem to seek credit merely for possessing a new technology, rather than using it to make better designs. Computers and their affiliated apparatus can do powerful things graphically, in part by turning out the hundreds of plots necessary for good data analysis. But at least a few computer graphics only evoke the response "Isn't it remarkable that the computer can be programmed to draw like that?" instead of "My, what interesting data." (Tufte, 2007)

There is, fortunately, an abundance of references in the areas of information visualization; in particular, *Envisioning Information* (Tufte, 1990) remains a classic both in the narrative but also on the fundamentals of displaying information.

So what can computers do for us in the areas of data visualization? The additional power of data visualization can be seen first by taking an example from *Semiology of Graphics* (Bertin, 2010). This example illustrates data clearly; however, when using a dot density version, it immediately draws the attention of the reader to the higher density areas. This would prompt the viewer to further investigate certain regions over others.

Perhaps the first attempt at recreating this in modern times using a computer would be the use of color, higher-definition graphics and precision. However, this adds little value to the overall chart, and would most likely induce confusion as to the meaning of the different colors where the eye is continuously shifting between legend and chart.

More interesting approaches to the use of computers would be the ability to display a view at a global scale or zooming into a particular region and breaking it down into smaller regions – geographic travel. Or perhaps, travel through time as the Trendalyzer software system, initially developed by the Gapminder Foundation (http://graphs.gapminder.org), demonstrates in a multivariable relationship bubble chart with the option to "travel" back and forth in time. This ability to visualize the changes between data time periods provides valuable information and insight which would have required a much higher investment in effort and would likely not be as effective in delivering the analysis.

The animation of moving charts, through time, geography, or any other dimension, must be added as a principal to computer graphic design. Animation, if used only to create an introductory effect, is distracting and will only contribute to the noise. Examples of bar charts growing into place one by one, or pie charts appearing in clockwise motion, delays the reading of the data and generates only enthusiasm for the computer's ability to perform these tasks. Having the data representation "react" to user interaction by means of an animation or displaying additional data is a much better use of this technology.

Consider one of the leaders of computer chart components, Dundas Data Visualization (www.dundas.com). Looking at the gallery of possibilities using their component, analysts become immersed in the ability of the component rather than the message that is being conveyed. Used effectively, the component can provide clear, detailed, interesting and valuable information with the possibility of highlighting specific areas of interest. Used incorrectly, and the analyst or application developer will spend a disproportionate amount of time attempting to understand the capabilities of the component.

One additional capability of computers in data visualization is their ability to record all interactions between analyst and data. If one was to view the history of the creation of a chart, additional information, or meta-information, could be obtained as to why this representation was chosen as the most appropriate display for the data by the user.

> When investigating data with visualizations, users regularly traverse the space of views in an iterative fashion. Exploratory analysis may result in a number of hypotheses, leading to multiple rounds of question-answering. Analysts can generate unexpected questions that may be investigated immediately or revisited later. After conducting analysis, users may need to review, summarize, and communicate their findings, often in the form of reports or presentations.
>
> By surfacing users' interaction history, we can facilitate analysis and communication. (Heer *et al.*, 2008)

Tableau software provides this capability by tracking historical data on the creation of the chart.

The insight gained in viewing this information, combined with the ability to communicate the story without having to be present, is a powerful tool that properly written software applications can provide.

Application algorithm and design

If the data visualization component is the heart of true CAIV, then the application is the brain. It is here that elements containing data from the hardware are transmitted to the output devices, where the interaction from the user via input devices is transformed in processing commands, and where the user can navigate and immerse in the world of visualization.

A lot can be learned from the electronic gaming industry in this area. After all, new generations will have the same expectations on navigation, image quality, speed, and even sound.

Business and scientific applications are not typically designed with strong user interfaces. In fact, most programmers, scientist and business ones alike, are rarely user interface or user experience experts, even more rarely having the capability to create high-quality computer graphics.

Take, for example, some of the most popular first-person games available on the market today. Their interfaces allow for the players to navigate in a 3D world easily using simple input devices such as a mouse and keyboard or a controller. They do that by constantly relaying information through dashboard types of display, but more importantly by allowing the player to orient themselves via the actual virtual landscape.

Applications should be designed so that users can orient themselves in their environment easily through understanding their relative positions; this would make thumbnail views simple as an additional navigation method. The user would be able to zoom in and out of the environment, or be able to discover functions within the screen to quickly understand how to manipulate the environment and its virtual objects. The user should naturally reach the intended results simply by moving to the virtual location.

Gears of War 2 has no visible command panels and navigation is done entirely based on the position of the player in relation to other objects and actors.

Learning from the movie industry, just as we did with their ability to imagine new input and output devices, the computer game industry has enriched their games with high-quality productions, sometimes more costly than actual movies, in order to increase engagement through visualization where the user is receiving massive amounts of information in seconds.

New generations of scientific and business analysts will have higher expectations in navigation and reaction times, comparing it many times to games they played in their teenage years (and perhaps still today). Today's application interfaces need to be redesigned to include many of the features available in first-person action games and real-life games.

Perhaps the greatest example of bringing life to an application is demonstrated by MASSIVE software (www.massivesoftware.com). Their motto of "Simulating

Life" perhaps explains its purpose, but their work can best be seen in films such as "i Robot" and "Lord of the Rings," among many others, as well as in countless television adverts. It is a software application that simulates, generates and renders crowd behaviors capable of autonomous and intelligent movement with personalized characteristics as well as reactive to their surroundings.

The creation of agents, complete with emotional qualities such as bravery or joy, each personalized and replicated in the hundreds and thousands, and rendered at such a high quality that a virtual eco system is created. This can typically be seen in most movie scenes where a large army of warriors fights another army (e.g., "Troy," "Lord of the Rings," "The Mummy," etc.), and yet the camera flying through the virtual world captures unique and individual behavior from each character.

The applications to computer/*in-silico* biological research here are staggering. Instead of emotional qualities, agents could be created with distinct physiological and disease characteristics. If an "army" of agents, each with slight variations, attack the army with assay after assay, we can monitor the reaction at the individual, the group, the culture, racial and disease levels. The possibilities are endless and applications and algorithms such as MASSIVE could be at the center of the new era of research.

Yet another critical part of the application design is the process by which the user finds information. Even if systems such as MASSIVE are capable of generating millions of possible outcomes and virtual environments, finding each possible outcome is still difficult. Today's search engines have moved from a categorization type of search (e.g., Yahoo) to a data element relationship-based search (e.g., Google). That means that we find information by locating similar information and narrowing it down based on connections to the final answer to the initial search starting point. Google calls this page rank, a ranking system that is based on how and how often pages reference each other.

The issue with this approach is the ability to find the initial starting point as close to the endpoint as possible. Recently, Google has added additional capabilities that suggest to the user words and phrases based on previous searches and indexed keywords. This is an attempt to assist the user with better formulating their search request and reaching the final search result faster.

If information is not cataloged and no searches have been done that can provide search metadata back to the search engine, searching continues to be a difficult problem to solve.

Wolfram|Alpha (www.wolframalpha.com) is attempting to solve this based on their extensive experience in computational mathematical problem-solving. Their product Mathematica was the product that spurred this new search engine to its current state. In Mathematica, a very generic symbolic language is available to create any type of model or system for any type of industry. In this language, not only can complex mathematical equations be developed and visualized, but there are functions that can represent outside knowledge. For example: ChemicalData["Caf feine,""MoleculeWeight"] results in 194.191 grams per mole.

In other words, in Mathematica an entire set of databases assist analysts by allowing them to focus on the calculations and research rather than looking for information. Simply by plugging in GenomeData, ProteinData, ChemicalData and many other types of data functions along with the required parameters, Mathematica retrieves the latest information directly from its internal databank.

Still, to use this capability one needs to know what to search for, what to plug in to the function. Wolfram|Alpha is proposing to solve this, interestingly enough, by using mathematics. By decomposing a search phrase and identifying those words which are data elements, those that are metric elements, a simple search such as "caffeine vs. aspirin structure" results in a side-by-side molecular structure diagram for both items.

Additionally, the introduction of curated data which bring data from some internal data sources while referencing hundreds of external references is a great step in ensuring that the latest data are used while providing additional research pathways into the original documents (research papers, reference material, databanks, etc.).

Tomorrow's search engines need to move towards a combination of all three types of searching methodologies. Simple categorization is still critical to ensure that basic information is easily retrieved, and the more complex relational ranking and searching metadata are highly effective in achieving the required speed. However, combining these features with a free-format searching engine, capable of identifying data and metric elements providing instant comparative analytical reports, will instead be based on categorical and search metadata, and will truly increase the efficiency of any software application.

User adaptation

Technology in its many forms can be complex and confusing, even for IT professionals. As computers grow more powerful and capabilities are ever-expanding, applications for technology exponentially increase, as does their complexity. Technology imposes changes in process and behavior and changes that can be difficult for people to accept. However, as Peter M. Senge states, "People don't resist change. They resist being changed" (Senge, 1994). The adoption of any new technology depends on the benefits to the users, whether they are tangible, intangible, or even emotional. Users typically have to adapt to a new technology and the steepness of the learning curve involved in adapting to a new technology can spell the difference between success and failure. User-centric design is very important for technology adaptation and how the technology fits the user's methods. CAIV is rapidly becoming a mainstream tool, as can be seen in television broadcasts, newspapers, and websites.

The user adoption of technology needs to provide clear benefits to the users, not just expected corporate return on investment and improvements to features, process, or reporting. IT projects have a history of failure when a new technology is imposed on a user base. Rapid technology adoption can happen, and a recent example is the iPod and iPhone from Apple. While the devices were innovative in their approach to providing technology for users, it was not the leading-edge technology available but how it was made available. Change management has a role to play in the implementation of technology projects and is really a project unto itself. Project failures can be blamed on many factors including improper, ineffective change management, but change management cannot manufacture user buy-in and engagement.

You can lead a horse to water but you can't make it drink, is a common proverb. The acceptance of change occurs when users are engaged and are vested in the process. Toyota adopted the Kaizen process of constant continual innovation that involves their workers instigating and managing manufacturing process improvements. The reasons that users accept a technology are not always the expected ones; for example, the Nokia N770 was panned by critics, but sold over a million units.

Historically, technology has been designed for technical users; even consumer products typically require a significant learning curve to properly use the software or even the basic functions. User-centric design is not just about simplifying the user interface, but making the software accessible to users by understanding the environment and how they use it. For software to be easy to use, it is important that its features are discoverable. It is common knowledge among software development professionals that software manuals and help files are the last resort of users, if they are used at all. Software needs to be designed to be simple, which usually takes a significant amount of thought and engineering, but not simplistic, which removes value from the software for users. Google is an example of user-centric design with its simple interface being a very deliberate choice, and efforts are ongoing in maintaining a very clean landing page even as the functionality provided by Google increases. With the increased processing power of computers, it is now easier to appeal to the visual nature of users, and even Google with its clean front-end interface has added iGoogle among many other applications which are graphically more demanding.

The brain is the most powerful pattern recognition system. CAIV is a tool to unleash that power. The user's familiarity with their current analysis tools will require adaptation to use CAIV; however, there are benefits in increasing visual elements for drug discovery, protein analysis, and basic research. The most important factor in the use of CAIV will be maintaining accessibility to users who are not highly technical so that the power of CAIV can be used by many.

It is here that IT departments supporting research and business facilities must also change their mindset. In their attempts at securing internal information and standardizing their world to make it easier to manage complexity, they sometimes undermine the end user with lower-quality applications, non-innovative networks, inferior hardware, and application barriers designed to keep intruders out but end up restricting internal users within a very small data world.

This is perhaps the biggest challenge to IT worldwide. While the outside world is opening its doors to external participants, most corporations (especially larger ones) are closing their doors to this readily available and cheap information.

Some examples of larger companies that have recently opened up their closed systems and are experiencing amazing success are: Microsoft with their open-source and Azure initiatives; Apple with their iPhone/iPod development partnership program; Amazon with their cloud services; and also in general all Open Source projects including Procter & Gamble's Innovation Network (www.scienceinthebox.com) program for scientific research in consumer products.

Companies that hold back on upgrading to the latest operating system because it is not stable, because it is too different from their current systems, because in their view the cost to benefit ratio is low, are fooling themselves in several ways. Firstly, eventually the upgrades will be required and with a gap between upgrades this will introduce a much higher cost. Secondly, the function of a technology department is to perform continuous improvements in the hardware and software infrastructure as well as to train their users. Delaying constant adoption and training on new technology is a detrimental process to the role. Finally, holding back upgrades (be they software or hardware) causes potential and current employees, partners and customers to regard their company as antiquated and not up-to-date. There is nothing that inspires less confidence than being the only conference participant with an outdated 10-kg laptop where your competitors have the latest, fastest and visually appealing system.

Security is seen as a sacred word to IT departments as it justifies the numerous firewalls, closed systems, locks and sometimes the inability to even select the desktop background of your choice. Firstly, security can be accomplished smartly by truly analyzing the entry and exit points of a network and not applying a "one rule fits all" scheme. Secondly, true security is impossible to achieve, and organizations that pride themselves in having the greatest security system enabled find themselves in dire straits when they are broken into, not only from the damage but from their high investment in securing the network which provided little return on investment.

IT needs to take a more customer conscious approach and deliver tailored solutions to different internal departments, groups, and people. They need to continuously invest in proper and sophisticated technology, and research technology that will give their organization a competitive edge rather than holding back the organization in stale security, procedures and outdated equipment and applications. For those items, there are plenty of outsource vendors ready to support organizations with those services, but there are very few people, other than IT people in an organization, that truly have a 360-degree view of their business and can truly assist with innovation, training and improvements on how to best use new technology.

Perhaps the above changes represent a harsh view of IT departments and is meant more as a general approach seen in the industry (particularly in the health industry) rather than a certitude. However, the *raison d'être* of the current state is not by choice but rather by the failure of the business owner or management to understand that if the strategy of the company is to be a leader on the scientific and business front a certain degree of investment, risk and change in IT is also part of the game.

In a finance-/legal-driven company, such change as described in most of the chapters is seen as risky, with little value to the organization. This is of course more apparent in shareholder/publicly held companies where long-term investments in research not directly related to the line of business (such as *in-silico* biology and other areas requiring strong IT investment) are seen as not profitable. Strong understanding, or otherwise belief, on the benefit of technology is needed in upper management to ensure breakthroughs in drug discovery.

Conclusion

To properly achieve optimal and effective information visualization, especially in computer/*in-silico* biological research, five key components are required for the researcher/data analyst. These five components are critical for effective and true CAIV, which is the next generation of data analysis and intelligence.

Additionally, data visualization, although critical to the overall CAIV process, is not the same as information visualization, and is one of the five components alongside input/output hardware, increasingly powerful computational hardware, strong application design and a change in the mindset of how organizations deal with investment and using future CAIV systems.

After some years of low growth in many of the above areas, much exists today in support of CAIV and much more is being developed on a daily basis. Research companies are slowly supporting this growth by dedicating substantial resources to IT innovation and support departments within their organization.

As a final comment, this chapter is merely a footnote in such an extensive subject. It is the hope that enough references were given for further research on this topic, and more importantly the realization that it is not enough to be able to produce data visualization, but rather to provide an environment where the researcher can become immersed and navigate, where information is acquired, manipulated, heard, and interacted with in real time. Only then will researchers and analysts focus on drug discovery rather than data manipulation.

References

Bertin, J. *Semiology of Graphics: Diagrams, Networks, Maps*. Redlands, CA: ESRI Press, 2010.

Boyer, M., Tarjan, D., Acton, S. and Skadron, K. Accelerating leukocyte tracking using CUDA: A case study in leveraging manycore coprocessors. In *Parallel and Distributed Processing Symposium, International* (pp. 1–12), 2009.

Fry, J. *Organic Information Design* (p. 14). Pittsbirgh, PA: Carnegie Mellon University, 1997.

Göbel, M., Müller, H. and Urban, B. *Visualization in Scientific Computing*. Vienna: Springer-Verlag, 2005.

Heer, J., Mackinlay, J., Stolte, C. and Agrawala M. *Graphical Histories for Visualization: Supporting Analysis, Communication, and Evaluation*. Tableau Software Whitepaper, 2008.

Lee, J. and Hudson, S. Foldable Interactive Displays.

Post, F., Nielson, G., Bonneau, G.-P. (Eds.) *Data Visualization: The State of the Art*. Dordrecht: Kluwer, 2003.

Senge, P. M. *The Fifth Discipline: The Art & Practice of The Learning Organization*. London: Doubleday Business Press, 1994.

Soukup, T. and Davidson, I. *Visual Data Mining – Techniques and Tools for Data Visualization and Mining*. Chichester: John Wiley & Sons, 2002.

Tufte, E. R. *Envisioning Information*. Cheshire, CT: Graphics Press, 1990.

The Visual Display of Quantitative Information (2nd edition). Cheshire, CT: Graphics Press LLC, 2007.

Yin, L. *3D Reconstruction and Virtual Reality Application in Financial Education and Financial Analysis*. Arts et Metiers ParisTech – Institut Image Le2i, France, 2014.

8 Example of computational biology at the new drug application (NDA) and regulatory approval stages: A case study on the comparative secondary pharmacologic analysis of the CCR5 antagonists maraviroc, vicriviroc, and aplaviroc

William T. Loging, Marilyn Lewis, Bryn Williams-Jones, and Roy Mansfield

Target-based prejudices

In the early 2000s, a popular and clinically valuable class of drugs known as the nonsteroidal anti-inflammatory drug (NSAID) class was being used to treat osteoarthritis and acute pain conditions. Of the several chemical members of this class, rofecoxib was approved by the Food and Drug Administration (FDA) in May 1999, with Merck and Co. marketing it under the brand names Vioxx and Ceoxx. The drug recorded more than $2 billion a year with more than 75 million patients taking it worldwide.[1]

In September 2004, rofecoxib was withdrawn from the market by Merck and Co. due to worries about its usage leading to increased risk of heart attack and stroke. This observation was noted in patients with long-term, high-dosage use. It has been estimated that these side effects led to the manifestation of heart disease in 100,000 cases.[2]

The FDA intently scrutinized other NSAIDs and continued to do so.[3] Several calls were made by concerned citizens and those affected by the rofecoxib side-effect case that the FDA needed to do more to insure public safety of post-approved drugs. Investigators began focusing on the target (known as COX-2), and it was hypothesized that the side effects observed were related directly to COX-2 inhibition. No

Bioinformatics and Computational Biology in Drug Discovery and Development, ed. W.T. Loging. Published by Cambridge University Press.

concrete data have been observed about whether the cardiovascular effects were solely related to COX-2. However, the mindset was already created that if a side effect was observed through certain drug use then all mechanisms of physiological effects were termed again the "Vioxx effect." This viewpoint – that all side effects and physiological effects were connected via a central target – impacted multiple other drug discovery projects as well, including the race to deliver the first CCR5 antagonists, which are a class of oral antiretroviral drugs designed to block HIV entry into cells via the CCR5 receptor. There were concerns that the novel mechanism of action of CCR5 antagonists was associated with potential risks of hepatotoxicity and malignancies, which might be adverse events that extend to all compounds in the class. The following work addressed such concerns by employing a wide range of computational approaches, utilizing pharmacology and human gene expression as well as clinical data to address these potential CCR5 antagonist class effects. Through investigating the pharmacologic properties between aplaviroc, vicriviroc, and maraviroc, we were able to show that all three compounds bind differently across a standard panel of assays. Collectively, multi-pharmacologic analysis and human gene expression data strongly suggest that there was no CCR5 antagonist-class relationship with hepatotoxicity or malignancy, and that maraviroc at approved therapeutic dosages is pharmacologically distinct from other members of this class.

Background on CCR5

The first step in the life cycle of human immunodeficiency virus 1 (HIV-1) is attachment to the CD4 cell surface receptor, but binding to this receptor alone is not sufficient to allow the virus to enter the cell. Binding to one of two chemokine receptors, CCR5 or CXCR4, is also required to allow infection to proceed.[4–6] In primary infection and earlier stages of disease, HIV-1 virus populations predominantly use CCR5, and 50%–62% of heavily antiretroviral treatment-experienced patients continue to harbor only CCR5-tropic virus (see Figure 8.1).[7–9]

Following attachment of the virus to the co-receptor, a cascade mechanism leads to virus–cell membrane fusion and release of the HIV genome into the cell.[10] The inhibition of the virus–cell fusion process has become an attractive target for antiretroviral drug development. CCR5 antagonists are a class of oral antiretroviral drugs designed to block HIV entry into cells via the CCR5 receptor.[11–13] The mechanism of action of CCR5 antagonists differs from other available antiretroviral drugs as they are the first approved drug class to target a human rather than a viral protein by binding to the host-cell receptor. To date, three CCR5 antagonists have entered clinical trials, namely aplaviroc, vicriviroc, and maraviroc,[14–16] with maraviroc the first of these to receive marketing approval.

Concerns have been raised over the safety of CCR5 antagonists following reports of hepatotoxicity and malignancy (particularly lymphomas) in trials of aplaviroc and vicriviroc, respectively.[17–21] In 2005, Phase II/III clinical trials of

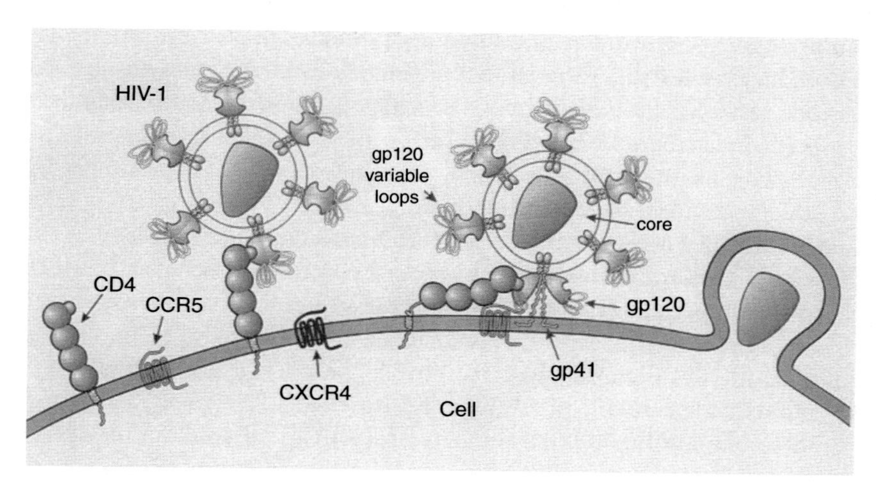

Figure 8.1 Attachment of HIV requires the gp120 loop to the CD4+ receptor. Upon attachment to a co-receptor (CCR5 or CXCR4), the invading HIV virion enters the cell. *Source*: Wikimedia Commons, PD-USGov-NIH.

aplaviroc were terminated in treatment-naive patients due to severe but reversible liver toxicity in one patient and asymptomatic rises in liver enzyme levels in several other patients.[19] Trials in treatment-experienced patients receiving aplaviroc continued, but were terminated when several patients developed severe hepatotoxicity.[19] The potential importance of CCR5 in preventing hepatotoxicity was supported by experiments in CCR5-deficient mice, which showed that administration of concanavalin A (an inducer of experimental T-cell-mediated hepatitis) led to immune-mediated liver disease and fulminant hepatitis.[22,23] Severe drug-related hepatotoxicity has also occurred in two patients enrolled in clinical trials of maraviroc,[24] but both cases were confounded by the administration of hepatotoxic drugs and, in one case, by bacterial sepsis. In a Phase II clinical trial of vicriviroc (ACTG-5211), six treatment-experienced patients who received vicriviroc developed a diverse set of malignancies, including lymphomas, compared to two patients with malignancies – both carcinomas – in the control group.[21] Taken together, these events, along with the novel mechanism of action of CCR5 antagonists, have led to concern that hepatotoxicity and/or malignancies may represent potential adverse event risks that extend to all compounds in the class.[18]

We set out to employ a wide range of computational approaches, utilizing pharmacology and human gene expression as well as clinical results data, to address the potential risks of CCR5 antagonists. We also investigated CCR5 tissue-specific gene expression, encompassing 90 normal human tissues across 13 organ systems, in order to identify tissues of interest that express CCR5 and might elicit a phenotypic change in response to a CCR5 antagonist.

In addition to binding to its primary protein target, many drugs can bind to other proteins within or even outside those associated with their drug families. The types and functions of these "non-targeted" proteins can play a role in the process of secondary phenotypes. Therefore, it is important to elucidate not just single proteins but panels of proteins in how a specific drug or even drug candidate may bind to them. Once the activities of a number of drug molecules are assessed across a panel of proteins, the molecules can be grouped together based on their properties and differences identified in such a panel.

Published work in the area of comparative pharmacology[25–29] suggests that highly similar binding profiles between compounds across a panel of assays can correlate with additional secondary target binding affinities and similar secondary physiologic *in vivo* effects. It has also been observed that compounds that have highly dissimilar activity profiles across an assay panel can have highly unrelated secondary physiologic *in vivo* effects. The process by which investigators compare the highly similar pharmacology of small molecules is referred to as biospectra analysis (sometimes referred to as chemogenomics or the BioPrint® approach).

As described above, biospectra analysis is conducted by screening small molecules at single high concentrations (10 μM) in a battery of bioassays representing a cross-section of the druggable proteome.[26] The percent inhibition values (biospectra) generated in such a bioassay array define the properties of the molecule and its ability to interact with the protein panel.[26] The extent of similarity in binding between a small molecule and other compounds in the data set is defined by the confidence in cluster similarity (CCS) scale, based on hierarchical clustering.[27] CCS values range from -1 (completely dissimilar) to $+1$ (identical). Compounds with CCS values of ≥ 0.7 are considered to be highly similar and therefore should be reviewed for their potentiality of having related pharmacologic properties;[30] this provides an unbiased novel tool for forecasting structure–response relationships and for translating broad biologic effect information into chemical structure design.[26] Similarity in the biospectra between compounds implies that these compounds will modulate similar protein pathways, and thus have similar biologic effects. Biospectra analysis approaches to comparative pharmacology have linked compounds with similar pharmacologic function, and have also dissociated compounds with distinct function, for example, in identifying agonists and antagonists of dopamine, α_2-adrenergic receptors, and serotonin 5HT1A.[26] Quantitatively grouping proteins based on the chemical similarity of their ligands not only reveals biologically sensible clusters, but also links between unexpected targets.[31] For instance, this approach divulged that methadone, emetine, and loperamide may antagonize muscarinic M3, α_2-adrenergic, and neurokinin NK2 receptors, respectively.[31] These predictions were subsequently confirmed experimentally. This type of analysis can be extended to assist in successfully forecasting compounds with similar adverse events.[25,30]

We conducted an analysis to compare the pharmacologic properties of aplaviroc, vicriviroc, and maraviroc across 41 standard assays (from the Cerep BioPrint® pharmacology panel[28]) in order to examine whether these novel CCR5 antagonists

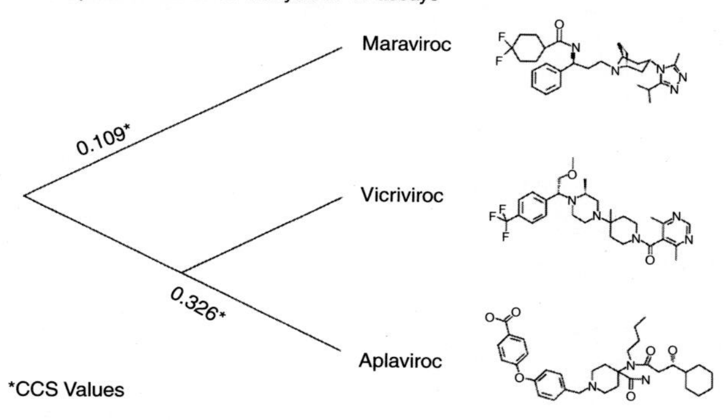

Figure 8.2 Dendrogram of the pharmacologic relationship of maraviroc, vicriviroc, and aplaviroc based on analysis of 41 assays. Pharmacology analysis based on comparisons of maraviroc, vicriviroc, and aplaviroc. The structure of each compound is shown (right). The dendrogram distances show that aplaviroc and vicriviroc are more related to each other pharmacologically than either of them are to maraviroc. This highlights the fact that maraviroc is both structurally as well as pharmacologically different from the other CCR5 antagonists.

were similar in their assay interaction, and therefore infer whether they would have similar secondary pharmacologic as well as phenotypic effects. In order to add further strength to our analysis, we also conducted human gene expression experiments and comparisons to known clinical results to address similarities and differences between the three CCR5 antagonists, as well as to assess CCR5 itself as a valid target.

The use of biospectra in comparison CCR5 antagonists

Our first goal was to investigate the pharmacologic properties of the CCR5 antagonists aplaviroc, vicriviroc, and maraviroc across a set of 41 standard assays in the BioPrint® database.[28] Nearest neighbors were analyzed using approaches similar to those utilized in other analyses that use highly dimensional data.[32] From the analysis, a dendrogram was constructed to assess the relatedness of each compound, with branch lengths dependent on cosine similarity (the smaller the CCS value, the longer the branch). The dendrogram constructed from the comparison between CCR5 antagonists is shown in Figure 8.2, with the chemical structures of the three antagonists shown alongside. Although maraviroc, vicriviroc, and aplaviroc are linear molecules of approximately the same size (molecular weights of 514, 533, and 578, respectively), they are not considered to be of the same chemical class and have no common functional groups (Figure 8.2). Significantly, aplaviroc has an acidic side chain whereas maraviroc does not. The core of maraviroc is a

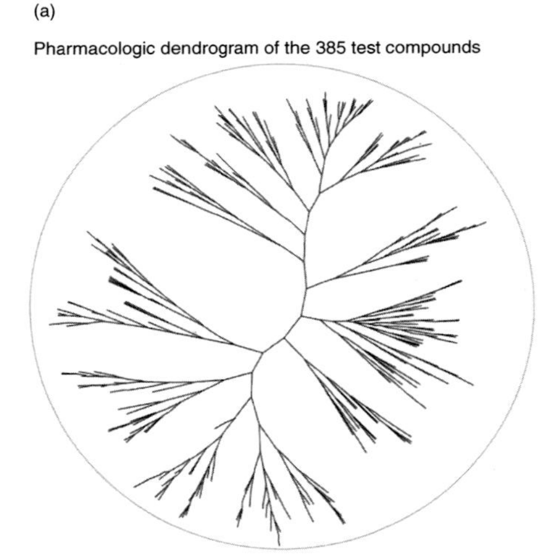

Figure 8.3a Pharmacologic dendrogram of the 385 test compounds. Phylogenetic tree of small-molecule pharmacology across 385 compounds, of which 382 are marketed drugs with the closest pharmacologic fingerprint to maraviroc, vicriviroc, and aplaviroc.

tropane, whereas for aplaviroc it is a spirocyclic diketopiperazine and for vicriviroc, a piperidino-piperazine. The pharmacology of both aplaviroc and vicriviroc were more closely related in our analysis, with a CCS value of 0.326, while maraviroc had a CCS value of 0.109 to the center node of the dendrogram, indicating a relatively distant relationship.

Next we assessed if any of the three CCR5 antagonists have similar pharmacology to other drugs on the market. Of more than 1000 known marketed drugs, we focused on a large cluster into which all three CCR5 antagonists fell, along with 382 other drugs. The resulting dendrogram reflects the diversity of the compounds tested (Figure 8.3a). For positive controls, three anti-fungal agents (clotrimazole, tioconazole, and miconazole) with closely related core structures and similar pharmacologic profiles (including a shared mechanism of action and similar adverse effects) were included in the analysis (Figure 8.3b). All three anti-fungal agents clustered closely together on a common branch of the dendrogram, with a CCS value of 0.725 – as would be expected from compounds that have related pharmacologic properties. Upon closer inspection of maraviroc, vicriviroc, and aplaviroc, all three compounds were noted to be located distally from each other on separate branches of the phylogenetic tree (Figure 8.4), indicating unrelated pharmacologic properties. Furthermore, this broader analysis yielded CCS values that were identical to those generated in the initial comparison between CCR5 antagonists, further supporting the likely pharmacologic disparity of these agents (Figure 8.2 and Figure 8.4).

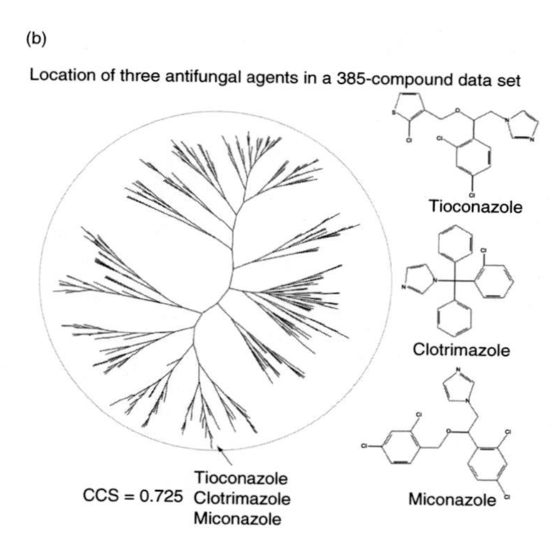

Figure 8.3b Location of three antifungal agents in a 385-compound data set. In order to access the association of other small molecules within this data set based on structure, three anti-fungal medicines (tioconazole, clotrimazole, and miconazole) were identified as positive controls. These three agents were identified as being near neighbors and clustered together on an arm of the tree.

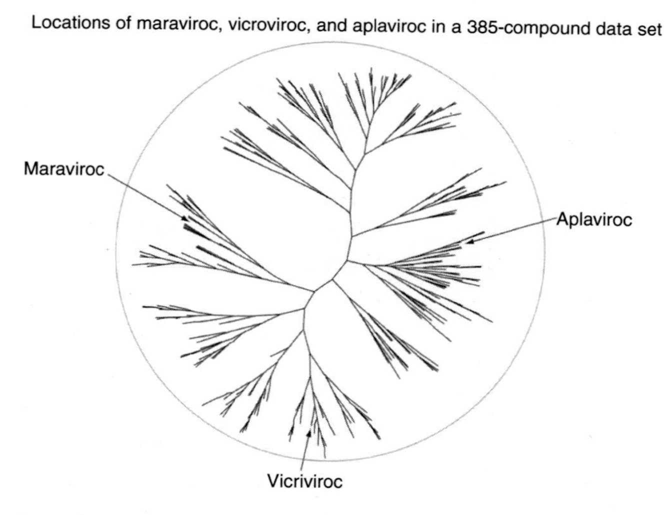

Figure 8.4 Locations of maraviroc, vicriviroc, and aplaviroc in a 385-compound data set. Pharmacology analysis of 385 small molecules, including maraviroc, vicriviroc, and aplaviroc. Each of these test compounds clustered distantly from each other within the phylogenetic tree, highlighting the fact that not only are maraviroc, vicriviroc, and aplaviroc structurally different from each other, but more importantly, they also differ pharmacologically in respect to the way they interact with the assay panel (biospectra).

Investigation of nearest neighbors for each of the three CCR5 antagonists

In the context of the other 382 drugs included in the assessment, maraviroc and its nearest neighbor had a CCS value of 0.45, which is below the threshold value of 0.7 that could indicate pharmacologic relatedness. This indicates that maraviroc does not have pharmacologic similarity to any of the marketed drugs in this panel. Similarly, aplaviroc and its closest neighbor had a CCS value of 0.46, which is not predictive of a significant pharmacologic similarity. Lopinavir was the nearest neighbor to vicriviroc, with a CCS value of 0.7. This may be reflected in adverse effects associated with these agents, given that gastrointestinal side effects are the main adverse events associated with lopinavir (Kaletra® Prescribing Information, USA, 2007) and abdominal pain was among the most common adverse events reported in patients treated with short-term vicriviroc monotherapy.[16,33]

Human tissue body map

In order to investigate whether CCR5 antagonists could directly cause hepatotoxicity by blocking CCR5 receptors on the surface of hepatocytes, we generated CCR5 gene expression data based on an analysis of 60 pathologically normal samples of human liver tissue derived from both male and female subjects. In order to properly gauge the gene expression of CCR5 in the context of the entire human body, we additionally generated a "whole body map" of CCR5 expression across more than 2000 tissue samples obtained from 90 healthy human tissues in 13 different organ systems. This analysis showed that the highest expression of CCR5 was on lymphocytes, monocytes, and spleen tissue, with no detectable expression of CCR5 in the liver (Figure 8.5), in contrast to other immune-related proteins that were expressed in the liver, such as complement component 5a receptor or chemokine (C-X-C motif) receptor 4 (CXCR4; data not shown; verified using Pfizer Body-Map system).

Bringing all of these data together for a more complete understanding of CCR5 antagonists

In these experiments, we sought to understand and compare the pharmacologic properties and differences of maraviroc, vicriviroc, and aplaviroc, as well as compare their protein binding to patterns of protein binding for other marketed drugs. Although these agents are all CCR5 antagonists, they are not considered to be of the same chemical class because they have no common functional groups. Previous observations have shown that drugs with the highest biospectra similarity have the highest structural similarity.[26] The fact that high CCS values correlate with molecular structure similarity is important, as it provides the ability to predict structure

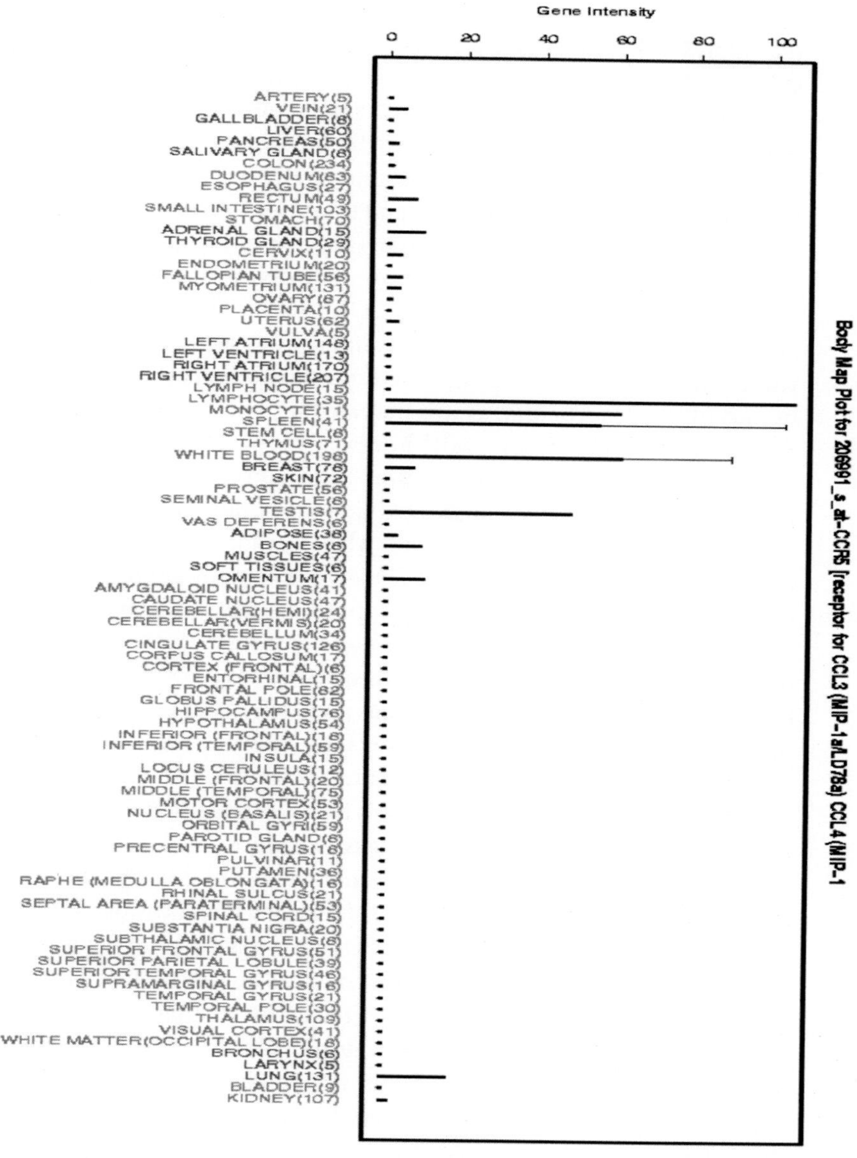

Figure 8.5 Body map showing expression of CCR5 in different body compartments and organs. Body map gene expression analysis conducted using Affymetrix GeneChip® technology in order to identify CCR5 expression across more than 2,000 tissue samples obtained using 90 whole pathologically normal human tissues from 13 different organ systems. The analysis showed the highest expression of CCR5 was on lymphocytes, monocytes, and spleen tissue, with no detectable expression of CCR5 in the liver (CCR5 gene expression based on an analysis of 60 normal samples of human liver tissue derived from both male and female subjects). Numbers in parentheses represent the number of tissue samples analyzed for each body compartment/organ. A black and white version of this figure will appear in some formats. For the color version, please refer to the color plate section.

from function, or function from structure.[27] Investigators have shown previously that high CCS values correlate with a similarity in the secondary pharmacologic effects of compounds.[25–27] Furthermore, biospectra analysis can be used to predict whether different compounds will have similar adverse events.[25,30] These researchers have extended this work to predict whether CCR5 antagonists have similar adverse event profiles, because a perception among HIV-treating physicians is that CCR5 antagonists are similar in structure and therefore will all be associated with hepatotoxicity and lymphomas as seen in trials of some of the individual compounds.[11,18,19]

Analysis against 41 standard assays included in the BioPrint® pharmacologic panel[28] demonstrated that the three CCR5 antagonists, maraviroc, vicriviroc, and aplaviroc, do not share a commonality in the way they bind a standard panel of proteins. Compounds with similar structure would be expected to cluster on the same branch of the dendrogram, as was the case for the three anti-fungal agents included as positive controls in our study; however, the three CCR5 antagonists that were studied were located on separate distal branches of the dendrogram. Differences in the chemical structures of aplaviroc, vicriviroc, and maraviroc, particularly variations in their side chains, may explain why these three agents were located on different branches of the dendrogram. The divergent positioning of CCR5 antagonists in our analyses argues against the hypothesis that long-term side effects will be comparable for all members of this class of drug. Further analysis showed that maraviroc and aplaviroc do not have a spectrum of secondary pharmacologic similarity to any of the 382 marketed drugs included in our analyses. The CCS value of the nearest neighbor to maraviroc was 0.45, which is below the threshold value of 0.7 that could indicate pharmacologic relatedness.[25–27]

Furthermore, the body map analysis demonstrated that CCR5 is not extensively expressed in the liver, and so hepatotoxicity due to the direct binding of maraviroc to CCR5 in the liver is unlikely to occur. The possibility remains that the hepatotoxicity seen in trials of aplaviroc was due to the direct action of the compound in ways other than direct binding to CCR5 on hepatocytes. This is not without precedent, as the preclinical development of at least one other CCR5 antagonist was halted due to hepatotoxicity caused by the direct action of the compound.[34] In this instance, steatosis induced in rats was suspected to have been a result of mitochondrial toxicity of the drug caused by the accumulation of endogenous ligands.

These findings are important because the antiretroviral treatment of HIV-infected individuals is long-term.[35,36] At least 1930 patients[37–39] have received maraviroc in clinical trials without an increased incidence of adverse events compared to control groups. In particular, hepatotoxicity, lymphoma, and other malignancies have occurred at a relatively similar frequency in patients receiving maraviroc compared with the frequency in patients receiving placebo.[24,40–42]

Data from the maraviroc clinical development program suggest that hepatotoxicity may not be a CCR5-class safety issue. Extensive evaluation of adverse event data from clinical trials of maraviroc revealed only sporadic liver enzyme abnormalities in the Phase I/IIa maraviroc program, with no dose relationship and no association with hyperbilirubinemia.[24] In the 48-week Phase IIb/III studies, there was no evidence of an imbalance of severe liver enzyme abnormalities or severe hepatic adverse events in maraviroc versus comparator arms, unadjusted for longer time on therapy in the maraviroc arms of the MOTIVATE studies.[24] The same finding was reported in HIV-infected patients co-infected with hepatitis B virus and/or hepatitis C virus,[24] who have an increased risk of hepatotoxicity.[43] Given the early occurrence of hepatotoxicity in the clinical evaluation of aplaviroc and the low frequency of elevated liver enzymes in individuals treated with maraviroc or vicriviroc versus comparator arms, even after 48 weeks of therapy, the possibility of hepatotoxicity being a class-specific adverse event is unlikely.[20]

In summary, comparative pharmacologic analysis of maraviroc, vicriviroc, and aplaviroc using a panel of 41 standard assays showed that these compounds do not share a common secondary pharmacologic biospectrum. While maraviroc and aplaviroc did not cluster closely with any of 382 marketed compounds, vicriviroc clustered with lopinavir. Collectively, biospectra analysis and the body map strongly suggested that there is no CCR5 antagonist class relationship with hepatotoxicity or malignancy, and that maraviroc at approved therapeutic dosages is pharmacologically distinct from other members of this class. Subsequently, additional clinical trails (MOTIVATE) have verified these analysis results through showing safety between maraviroc and placebo groups.[44]

Scientific methods used in this study

Protein panel comparisons to the druggable proteome

As described in earlier publications,[25–27] we set out to compare the protein-binding panel to the druggable proteome. The entire human proteome (some 60k sequences) was transformed into 700 consensus sequences that were related to specific druggable protein families. The assay panel was then mapped on top of this tree in order to assess their breadth of coverage across the proteome, which showed that the assay panel has a broad coverage across many disparate protein families.

Data analysis

BioPrint® database

The BioPrint® database includes a homogeneous set of data generated by testing an extensive range of marketed drugs, failed drugs, and reference compounds in a panel of well-characterized *in vitro* assays, including a diverse selection of molecular targets. It contains absorption, distribution, metabolism, and excretion data for the drugs in the data set, together with *in vivo* data related to therapeutic use, pharmacokinetics, and adverse reactions.[25] This allows for the development of *in vitro/in vivo* associations covering a range of clinical outcomes. At the time of our analysis, the database included information on binding activities for 1100 compounds against a panel of up to 122 assays, providing more than 125,000 data points.

We used data obtained for maraviroc across a set of 41 standard assays[26] (Supplementary Table 8.1) to compare percentage inhibition values to data available for aplaviroc and vicriviroc against the same assay set. The specificity of binding of the ligand to the receptors was defined as the difference between the total binding and the non-specific binding determined in the presence of excess unlabeled reference ligand (available at: www.cerep.fr/cerep/users/pages/downloads/general.asp). The results were expressed as a percent inhibition of control-specific binding (non-specific and total binding) obtained in the presence of maraviroc. A dose–response curve was calculated using an eight-point dose response of the relevant reference compound. The 50% inhibition concentration (IC_{50}; concentration causing a half-maximal inhibition of control-specific binding) values and Hill coefficients (n_H) were determined by non-linear regression analysis of the competition curves using Hill equation curve fitting. The inhibition constants (K_i) were calculated from the Cheng Prusoff equation ($K_i = IC_{50}/[1+\{L/K_D\}]$, where L = concentration of radioligand in the assay and K_D = affinity of the radioligand for the receptor). In the distance calculation (see below), J is defined as an observation or measurement within a profile. In each experiment, the respective reference compound was tested concurrently against maraviroc, vicriviroc, and aplaviroc in order to assess the suitability of the assay. Both the reference compounds and the CCR5 antagonists were tested at several concentrations (for IC_{50} value determination) and the values were compared to previously determined historical values in the Cerep database (available at: www.cerep.fr/cerep/users/pages/downloads/pharmacology.asp). The assay was rendered valid if the suitability criteria were met, in accordance with the corresponding operating procedure outlined on the website.

Calculation of CCS values

Correlation scores (CCS values) were assigned to each compound based on how similar the profiles were to those with the nearest likeness, or "nearest neighbor." This value ranges from –1 (completely dissimilar) to +1 (completely identical), and is calculated using the cosine correlation measure; for two profiles, a and b, with k dimensions, this calculation is: $\frac{\sum_{j=1}^{k} a_j b_j}{\text{norm}(a)\text{norm}(b)}$ where, $\text{norm}(a) = \sqrt{\sum_{j=1}^{k} a_j^2}$

Construction of dendrograms

From the CCS values, a dendrogram[45] can be constructed to visually inspect the relatedness of compounds; if a pair of compounds are linked by a branch, then

Supplementary Table 8.1 List of the 41 assays used in the biospectrum analysis.

Assay description
5-HT transporter human
5-HT1A human
5-HT1B
5-HT2A human
Acetylcholinesterase human
Adenosine 1 human
Adenosine 3 human
Adrenergic alpha 1 non-selective
Adrenergic alpha 2A human
Adrenergic alpha 2B
Adrenergic alpha 2C human
Adrenergic beta 1 human
Adrenergic beta 2 human
Angiotensin-converting enzyme human (recombinant enzyme)
BZD (central)
CA2+ channel (L, diltiazem site) (benzothiazepines)
CA2+ channel (L, verapamil site) (phenylalkylamines)
Choline transporter
Dopamine D1 human
Dopamine D2 human
Dopamine transporter human
Endothelin A human
Endothelin B human
GABA transporter
Ghrelin human
Glutamate receptor NMDA site
Histamine 1 central
Histamine 2
L-type calcium channel, DHP site
MAO-A human
Muscarinic 1 human
Muscarinic 2 human
Muscarinic 3 human
N (neuronal) (alpha-BGTX-insensitive) (alpha 4 beta 2)
Na/K ATPase enzyme activity
Norepinephrine transporter human
Phosphodiesterase 3 human
Phosphodiesterase 4 human

Supplementary Table 8.1 (*cont.*)

Assay description
Phosphodiesterase 5 human
Urotensin 2 binding to UT1 human receptor
Vasopressin 1A human

they would be expected to have related pharmacologic properties. Clustering was performed as previously published[27] with Spotfire software, using the UPGMA (unweighted pair-group method with arithmetic mean) algorithm and inputting the CCS values as the similarity measurements. Phylogenetic trees were then generated based on these clusters using the HyperTree software tool,[45] which displays the phylogenetic relatedness of compounds via one of three styles of graphical output: a dendrogram, a radial view, or a hyperbolic view.[45]

Expanding the data set

In order to expand this work and compare the three CCR5 antagonists to a larger set of compounds, we conducted the same experiment (as described above for maraviroc, vicriviroc, and aplaviroc) with an additional 382 marketed drugs, bringing the total data set to 385 compounds. We investigated if any other drugs could be identified as having similar pharmacologic properties to the three CCR5 antagonists.

Body map

The CCR5 gene expression analysis was conducted using Human Genome U133A/B GeneChip® arrays with standard Affymetrix protocols (www.affymetrix.com). All of the data were normalized into a single expression file containing averages and mean absolute deviations across each of the human tissue types. Normalization of the file was performed using the MAS5 Statistical Algorithm.

References

1. Oberholzer-Gee F, Inamdar SN. Merck's recall of rofecoxib – A strategic perspective. *New England Journal of Medicine*. 2004;351(21):2147–2149.
2. Bhattacharya, S. Up to 140,000 heart attacks linked to Vioxx. *New Scientist*. 2005, Jan 25.
3. www.fiercepharma.com/special-reports/top-10-pharma-settlements/merck-vioxx
4. Alkhatib, G., Combadiere, C., Broder, C. C., *et al.* CC CKR5: A RANTES, MIP-1alpha, MIP-1beta receptor as a fusion cofactor for macrophage-tropic HIV-1. *Science*. 1996;272:1955–1958.
5. Dragic, T., Litwin, V., Allaway, G. P., *et al.* HIV-1 entry into CD4+ cells is mediated by the chemokine receptor CC-CKR-5. *Nature*. 1996;381:667–673.
6. Feng, Y., Broder, C. C., Kennedy, P. E. and Berger, E. A. HIV-1 entry cofactor: Functional cDNA cloning of a seven-transmembrane, G protein-coupled receptor. *Science*. 1996;272: 872–877.

7. Wilkin, T. J., Su, Z., Kuritzkes, D. R., *et al.* HIV type 1 chemokine coreceptor use among antiretroviral-experienced patients screened for a clinical trial of a CCR5 inhibitor: AIDS Clinical Trial Group A5211. *Clinical Infectious Diseases*. 2007;44:591–595.
8. Coakley, E., *et al.* 2nd International Workshop on Targeting HIV Entry. Boston, USA, October 20–21, 2006.
9. Whitcomb, J. M., *et al.* 10th Conference on Retroviruses and Opportunistic Infections USA. Boston, USA, February 10–14, 2003.
10. Moore, J. P. and Doms, R. W. The entry of entry inhibitors: A fusion of science and medicine. *Proceedings of the National Academy of Sciences, USA*. 2003;100:10598–10602.
11. Idemyor, V. Human immunodeficiency virus (HIV) entry inhibitors (CCR5 specific blockers) in development: Are they the next novel therapies? *HIV Clinical Trials*. 2005;6:272–277.
12. Lederman, M. M., Penn-Nicholson, A., Cho, M. and Mosier, D. Biology of CCR5 and its role in HIV infection and treatment. *Journal of the American Medical Association*. 2006;296:815–826.
13. Westby, M. and van der Ryst, E. CCR5 antagonists: Host-targeted antivirals for the treatment of HIV infection. *Antiviral Chemistry and Chemotherapy*. 2005;16:339–354.
14. Fätkenheuer, G., Pozniak, A. L., Johnson, M. A., *et al.* Efficacy of short-term monotherapy with maraviroc, a new CCR5 antagonist, in patients infected with HIV-1. *Nature Medicine*. 2005;11:1170–1172.
15. Lalezari, J., Thompson, M., Kumar, P., *et al.* Antiviral activity and safety of 873140, a novel CCR5 antagonist, during short-term monotherapy in HIV-infected adults. *AIDS*. 2005;19:1443–1448.
16. Schurmann, D., Fätkenheuer, G., Reynes, J., *et al.* Antiviral activity, pharmacokinetics and safety of vicriviroc, an oral CCR5 antagonist, during 14-day monotherapy in HIV-infected adults. *AIDS*. 2007;21:1293–1299.
17. Antiviral Agents. Vicriviroc: Is the risk of cancer increased? *Treatment Update*. 2006;18:5–6.
18. Clotet, B. CCR5 inhibitors: Promising yet challenging. *Journal of Infectious Diseases*. 2007;196:178–180.
19. Deeks, S. G. Challenges of developing R5 inhibitors in antiretroviral naive HIV-infected patients. *Lancet*. 2006;367:711–713.
20. Emmelkamp, J. M. and Rockstroh, J. CCR5 antagonists: Comparison of efficacy, side effects, pharmacokinetics and interactions – Review of the Literature. *European Journal of Medical Research*. 2007;12:409–417.
21. Gulick, R. M., Su, Z., Flexner, C., *et al.* Phase 2 study of the safety and efficacy of vicriviroc, a CCR5 inhibitor, in HIV-1-infected, treatment-experienced patients: AIDS clinical trials group 5211. *Journal of Infectious Diseases*. 2007;196:304–312.
22. Ajuebor, M. N., Aspinall, A. I., Zhou, F., *et al.* Lack of chemokine receptor CCR5 promotes murine fulminant liver failure by preventing the apoptosis of activated CD1d-restricted NKT cells. *Journal of Immunology*. 2005;174:8027–8037.
23. Moreno, C., Gustot, T., Nicaise, C., *et al.* CCR5 deficiency exacerbates T-cell-mediated hepatitis in mice. *Hepatology*. 2005;42:854–862.
24. Hoepelman, I. M., *et al.* Presentation LBP7.9/1, 11th European AIDS Conference. Madrid, Spain, October 24–27, 2007.
25. Fliri, A. F., Loging, W. T., Thadeio, P. F. and Volkmann, R. A. Analysis of drug-induced effect patterns to link structure and side effects of medicines. *Nature Chemical Biology*. 2005;1:389–397.

26. Fliri, A. F., Loging, W. T. and Volkmann, R. A. Cause–effect relationships in medicine: a protein network perspective *Discovery Medicine*. 2011;11(57):133–143.
27. Fliri, A. F., Loging, W. T., Thadeio, P. F. and Volkmann, R. A. Biological spectra analysis: Linking biological activity profiles to molecular structure. *Proceedings of the National Academy of Sciences, USA*. 2005;102:261–266.
28. Krejsa, C. M., Horvath, D., Rogalski, S. L., *et al.* Predicting ADME properties and side effects: The BioPrint approach. *Current Opinion in Drug Discovery & Development*. 2003;6:470–480.
29. Guba, W. Compound Profiling & Chemogenomic Approaches, Philadelphia, USA, October 11–12, 2007.
30. Campillos, M., Kuhn, M., Gavin, A. C., Jensen, L. J. and Bork, P. Drug target identification using side-effect similarity. *Science*. 2008;321:263–266.
31. Keiser, M. J., Roth, B. L., Armbruster, B. N., *et al.* Relating protein pharmacology by ligand chemistry. *Nature Biotechnology*. 2007;25:197–206.
32. Clarke, R., Ressom, H. W., Wang, A., *et al.* The properties of high-dimensional data spaces: Implications for exploring gene and protein expression data. *Nature Reviews Cancer*. 2008;8:37–49.
33. Zingman, B., *et al.* Poster 795. 15th Conference on Retroviruses and Opportunistic Infections. Boston, USA, February 3–6, 2008.
34. Cornwell, P. D. and Ulrich, R. G. Investigating the mechanistic basis for hepatic toxicity induced by an experimental chemokine receptor 5 (CCR5) antagonist using a compendium of gene expression profiles. *Toxicology and Pathology*. 2007;35:576–588.
35. Lohse, N., Hansen, A. B., Gerstoft, J. and Obel, N. Improved survival in HIV-infected persons: Consequences and perspectives. *Journal of Antimicrobial Chemotherapy*. 2007;60:461–463.
36. Hamers, F. F. and Downs, A. M. The changing face of the HIV epidemic in western Europe: What are the implications for public health policies? *Lancet*. 2004;364:83–94.
37. Pfizer Data on file; presented as part of the FDA Antiviral Drugs Advisory Committee meeting, held April 24, 2007. Available at: www.fda.gov/ohrms/dockets/ac/cder07.htm#AntiviralDrugs.
38. Abel, S., *et al.* Poster 8, 8th International Workshop on Clinical Pharmacology of HIV Therapy. Budapest, Hungary, April 16–18, 2007.
39. Abel, S., *et al.* Abstract 55, 8th International Workshop on Clinical Pharmacology of HIV Therapy. Budapest, Hungary, April 16–18, 2007.
40. Lalezari, J., *et al.* Presentation H-718a, 47th Interscience Conference on Antimicrobial Agents and Chemotherapy. Chicago, USA, September 17–20, 2007.
41. Fätkenheuer, G., *et al.* Presentation PS3/5. 11th European AIDS Conference. Madrid, Spain, October 24–27, 2007.
42. Goodrich, J. M., *et al.* Presentation LB-2, 45th Annual Meeting of Infectious Diseases Society of America. San Diego, USA, October 4–7, 2007.
43. Sulkowski, M. S., Thomas, D. L., Chaisson, R. E. and Moore, R. D. Hepatotoxicity associated with antiretroviral therapy in adults infected with human immunodeficiency virus and the role of hepatitis C or B virus infection. *Journal of the American Medical Association*. 2000;283:74–80.
44. Stephenson, J. Researchers buoyed by novel HIV drugs. *Journal of the American Medical Association*. 2007;297(14).
45. Bingham, J. and Sudarsanam, S. Visualizing large hierarchical clusters in hyperbolic space. *Bioinformatics*. 2000;16:660–661.

9 Clinical trial failures and drug repositioning

Mark Crawford and Jeff Handler

More than 90% of compounds entering clinical trials fail to reach the market; thus, any company with a history in drug development has a large number of failed compounds "sitting on the shelf." Each of these compounds has a substantial portfolio of associated data and represents a significant sunk cost. This chapter discusses methods to find new uses for failed clinical-stage compounds (repositioning), and highlights a specific current example to discuss the unique development considerations for a repositioned compound.

Background

There is a long and rich history of successful drug repositioning. This includes repositioning of compounds during their first clinical development program, repositioning of compounds that failed in their original development program, and repositioning (or line extension) of marketed drugs. Selected examples are shown in Table 9.1. While the development strategies for new uses of marketed and failed compounds differ in regulatory, intellectual property and market protection considerations, the strategies for exploring for new uses are similar.

There has recently been a concerted effort in many pharmaceutical companies to systematically review all failed clinical candidates (FCC). This is driven by several considerations. First, there is general recognition that the highly focused approach necessary in the original development may have precluded consideration of all possible uses. Second, modern compound testing and data analysis paradigms are readily optimized for repositioning. Third, when a new development path is found, the development cost for an FCC could be significantly lower than the cost of *de novo* development of a new compound, as a large amount of proprietary non-clinical and clinical data exist for the FCC. Finally, as original composition of matter patents approach expiration, these FCCs become attractive targets for a range of

Bioinformatics and Computational Biology in Drug Discovery and Development, ed. W.T. Loging. Published by Cambridge University Press.

Table 9.1 Examples of repositioned drugs.

Drug	Original use	Reorientation
Amantadine	Influenza (DuPont)[a]	Parkinson's (DuPont)[a]
Atomoxetine	Depression (Lilly, not marketed)[b]	ADHD (Lilly)[b] Parkinson's, generalized social anxiety disorder, Huntington's disease, schizophrenia, cocaine dependence (Lilly, not marketed)[c]
AZT	Cancer (Michigan Cancer Foundation, not marketed)[d]	HIV (Burroughs-Wellcome)[e]
Bupropion	Depression (Burroughs Wellcome)[f]	Smoking cessation, seasonal affective disorder (GSK)[c]
Celecoxib	Arthritis (GD Searle)[g]	Familial adenomatous polyposis (Pfizer)[c]
Chlorpromazine	Anti-emetic (Rhône-Poulenc)[h]	Anti-psychotic (Smith-Kline & French)[h]
Dapoxetine	Depression (Lilly; not marketed)[i]	Premature ejaculation (J&J)[i]
Depo-Provera	Birth control (Pfizer)[j]	Endometriosis (Pfizer)[j]
Diphenhydramine	Antihistamine (generic)	Sleep aid >> anxiolytic (generics)
Duloxetine	Depression (Lilly)[k]	Generalized anxiety disorder[l], diabetic peripheral neuropathic pain[m], fibromyalgia[l], stress urinary incontinence[n] (Lilly) Multiple sclerosis pain, chronic low back pain, posttraumatic stress disorders, obsessive compulsive disorder, irritable bowel syndrome, chronic fatigue syndrome, osteoarthritis knee pain (Lilly; not marketed)[c]
Eflornithine	Cancer (Sanofi; not marketed)[o]	Anti-infective – trypanasomiasis (Sanofi)[o] Hirsutism (Shire; not marketed)[c] Bladder cancer (Genzyme; not marketed)[c]
Finasteride	Benign prostatic hyperplasia (Merck)[k]	Alopecia (Merck)[k]
Fluoxetine	Depression (Lilly)[k]	Bulimia nervosa[p], anorexia nervosa[p], panic disorder[p] and premenstrual dysphoric disorder[k] (Lilly) Autism (Neuropharm; not marketed)

Galantamine	Polio/paralysis/anesthesia[k] (Sopharma)	Alzheimer's (J&J)[k]
Glycopyrrolate	Ulcer, co-administered with anesthetics (generic)	COPD (Novartis, not marketed)
Lidocaine	Local anesthesia (AstraZeneca)[k]	Oral corticosteroid-dependent asthma (Coru, not marketed)[c] Dysmenorrhea (Columbia Laboratories, not marketed)[c] Premature ejaculation (Plethora Solutions, not marketed)[c]
Mecamylamine	Hypertension (Layton)[k]	S(+) enantiomer for depression (Targacept; not marketed)[c] Age-related macular degeneration, diabetic macular edema (CoMentis, not marketed)[c]
Mifepristone (RU487)	Pregnancy termination (Roussel Uclaf)[k]	Psychotic major depression, Cushing's syndrome, Alzheimer's disease (Corcept; not marketed)[c] Uterine fibroids (BioPro Medical; not marketed)[c] Hepatitis C (VGX, not marketed)[c]
Milnacipran	Depression (Pierre Fabre)[k]	Fibromyalgia syndrome (Cypress/ Forest)[q]
Minoxidil	Hypertension (Upjohn)[k]	Androgenic alopecia (Upjohn)[k]
Paclitaxel	Breast cancer (BMS)[k]	Restenosis (Boston Scientific)[k]
Phentolamine	Hypertension (Novartis)[k]	Night blindness (Ocularis; not marketed)[k] Restoration of feeling after dental anesthesia (Novalar)[c]
Premarin	Estrogen replacement (Wyeth)	Osteoporosis (Wyeth)
Raloxifene	Breast and prostate cancer (Lilly)[k]	Osteoporosis (Lilly)[k]
Retin-A	Acne (J&J)	Palliation of fine facial wrinkles (J&J) Acute promyelocytic leukemia (Roche) Cutaneous lesions in AIDS-related Kaposi's sarcoma (Ligand) Rosacea (Coria; not marketed)
Ropinirole	Hypertension (SKB)[k]	Parkinson's, restless leg syndrome (GSK)[k] Sexual dysfunction secondary to antidepressant pharmacotherapy, depression in bipolar disorder, fibromyalgia, major depression, motor recovery after stroke (GSK; not marketed)[c]

Table 9.1 (*cont.*)

Drug	Original use	Reorientation
Sibutramine	Depression (Boots)[k]	Obesity (Abbott)[k]
Sildenafil	Pulmonary hypertension (Pfizer)[k]	Erectile dysfunction[k], pulmonary arterial hypertension[r] (Pfizer) Female sexual arousal disorder, Meniere's disease, pre-eclampsia, chronic fatigue syndrome, idiopathic pulmonary fibrosis, chronic obstructive pulmonary disease, diabetic peripheral neuropathy (Pfizer; not marketed)[c]
Tadalafil	Inflammation/ hypertension (GSK; not marketed)[k]	Erectile dysfunction (Lilly/Icos)[k] Pulmonary arterial hypertension, dyspepsia, benign prostatic hyperplasia, Raynaud's phenomenon (Lilly; not marketed)[c]
Tetracyclines	Anti-infectives	Multiple sclerosis, spinal muscular atrophy (Paratek; not marketed)
Thalidomide	Sedative, anti-emetic (Grunenthal)[k]	Erythema nodosum leprosum, multiple myeloma (Celgene)[k] Melanoma, glioma, Waldenstrom's macroglobulinemia, idiopathic pulmonary fibrosis, amyotrophic lateral sclerosis, leukemia, HIV infection, uveitis, metastatic renal cell carcinoma (Celgene; not marketed)[c]
Tofisopam	Anxiety (EGIS)[k]	R-isomer for irritable bowel syndrome (Pharmos, not marketed)[k]
Topiramate	Anticonvulsant (McNeil)[k]	Migraine, post-traumatic stress disorder, obesity, pathological gambling, obsessive-compulsive disorder, restless legs syndrome, bipolar I disorder, cyclic vomiting syndrome, Tourette syndrome, smoking cessation, diabetic peripheral polyneuropathy, medication overuse headaches (J&J, not marketed)[c]

Notes: This is a selection of repositioned drugs with a bias towards corporate examples. There are a number of academic and institutional drug repositioning efforts that are not included. The comment "not marketed" can mean no longer in clinical development or currently in clinical development but not yet approved.

[a] David A. Hounshell, John Kenly Smith. *Science and Corporate Strategy: Du Pont R&D, 1902–1980* (1988) Cambridge University Press, p. 469. Accessed at: http://books.google.com/books?id=6ld0K9VNpmIC

[b] Melissa Garland and Peter Kirkpatrick. Fresh from the pipeline: Atomoxetine hydrochloride. *Nat Rev Drug Disc* (2004) 3; 385–386.

[c] www.clintrials.gov search using drug name with limitation of "industrial sponsor". Accessed April 17, 2009.

[d] A Failure Led to Drug Against AIDS. Published in the *New York Times* on September 20, 1986.

[e] MA Fischl, DD Richman, MH Grieco, MS Gottlieb, PA Volberding, OL Laskin, JM Leedom, JE Groopman, D Mildvan, RT Schooley, et al. The efficacy of azidothymidine (AZT) in the treatment of patients with AIDS and AIDS-related complex. A double-blind, placebo-controlled trial. *N Engl J Med* (1987) 317; 185–191.

[f] www.accessdata.fda.gov/scripts/cder/drugsatfda/index.cfm?fuseaction=Search.Label_ApprovalHistory#apphist Accessed April 17, 2009.

[g] www.accessdata.fda.gov/scripts/cder/drugsatfda/index.cfm?fuseaction=Search.DrugDetails Accessed April 17, 2009.

[h] Wikipedia contributors. Chlorpromazine. Wikipedia, The Free Encyclopedia. April 16, 2009, 17:09 UTC. Available at: http://en.wikipedia.org/w/index.php?title=Chlorpromazine&oldid=284239758. Accessed April 17, 2009.

[i] Wikipedia contributors. Dapoxetine. Wikipedia, The Free Encyclopedia. April 17, 2009, 14:59 UTC. Available at: http://en.wikipedia.org/w/index.php?title=Dapoxetine&oldid=284428899. Accessed April 17, 2009.

[j] Wikipedia contributors, 'Depo-Provera', Wikipedia, The Free Encyclopedia, 15 April 2009, 06:50 UTC, <http://en.wikipedia.org/w/index.php?title=Depo-Provera&oldid=283950598> [accessed 23 April 2009].

[k] Ted T Ashburn, Karl B Thor. Drug Repositioning: Identifying and developing new uses for existing drugs. *Nat Rev Drug Disc* (2004) 3; 673–683.

[l] www.fda.gov/cder/foi/label/2009/021427s021s027s028lbl.pdf

[m] Anathea B.Waitekus, Peter Kirkpatrick. Fresh from the pipeline: Duloxetine hydrochloride. *Nat Rev Drug Disc* (2004) 3; 907–908.

[n] http://emc.medicines.org.uk/medicine/14930/SPC/Yentreve++20mg+and+40mg+hard+gastro-resistant+capsules/

[o] Wikipedia contributors, 'Eflornithine', Wikipedia, The Free Encyclopedia, 20 April 2009, 19:40 UTC, <http://en.wikipedia.org/w/index.php?title=Eflornithine&oldid=285077484> [accessed 23 April 2009].

[p] http://dailymed.nlm.nih.gov/dailymed/drugInfo.cfm?id=6328#nlm34067-9

[q] http://dailymed.nlm.nih.gov/dailymed/drugInfo.cfm?id=9073#nlm34067-9

[r] http://dailymed.nlm.nih.gov/dailymed/drugInfo.cfm?id=9309#nlm34067-9

competing drug development companies. It would be difficult for any company to see a competitor turn their unwanted compounds into profitable products.

As can be seen in Table 9.1, most of the examples of drug repositioning have started from a currently marketed drug rather than from a FCC. The reasons for this are both economic and historical.

- It is significantly less expensive to get approval for the new use of a currently marketed compound than for a compound that has never reached the market.
- Marketed compounds can have a history of off-label use that indicates a potential new use.
- There has been significant historical effort within pharmaceutical companies at line extension for current products.
- As the original composition of matter patents come close to expiration, the significant amount of publicly available data on marketed compounds make them attractive repositioning targets for small companies.

Sorting through FCCs

Compounds fail in clinical development due to one or a combination of three considerations. Some compounds appear to be effective at the disease target but show unexpected toxicity, some show poor or no efficacy, and some are stopped based on market considerations. When compounds fail due to toxicity issues but act effectively on the disease target, several approaches (often in combination) have been successful in repositioning.

- Directly addressing toxicity through changes in formulation and route of administration.
- Addressing toxicity indirectly by choosing a different disease target affected by the same therapeutic mechanism. This implies that a narrow therapeutic index is acceptable in the new target disease (e.g., switching from rheumatoid arthritis to oncology), dosage can be significantly lowered, or concomitant reformulation/route of administration changes address toxicity issues.
- Exploiting a previously unappreciated property of the FCC that points to a different target disease where the toxicity can be better managed.

When compounds show poor or no efficacy but have a good safety profile, several approaches have been successful.

- If lack of efficacy is due to poor bioavailability, pro-drug development, reformulation and/or use of a new route of administration can be undertaken.
- If poor efficacy is due to inherently insufficient effect at the molecular target, a previously unappreciated property of the FCC may be exploited.
- Likewise, if poor efficacy is due to failure of the disease hypothesis (action at the expected molecular target does not ameliorate the disease), a previously unappreciated pharmacological property of the FCC may be explored (e.g., the use of sildenafil for treatment of erectile dysfunction as an unexpected indication).

When clinical development was stopped due to market considerations, there may be several opportunities.

- If development was stopped due to change in therapeutic area focus of the company, it may be possible to restart development in the original indication either in the originating company or in another company via outlicensing.
- If development was stopped due to the prediction that the compound would not be in a good competitive position relative to other marketed drugs, it may be possible to address the shortcomings through the methods listed above (reformulation or exploitation of a previously unappreciated property).

Reformulation strategies are beyond the scope of this chapter. Here we focus first on the discovery of an exploitable and previously unappreciated property of the FCC and second on development for a fundamentally new use.

A drug looking for a disease

Although there are not well-disclosed descriptions of recent FCC repositioning efforts, several are currently in development.[1–6] Looking at repositioned drugs in general (FCCs and those repositioned from currently marketed drugs), many had their origin as the result of a unique insight coming from one or a small group of people. The genesis of such insights is not well documented, but likely comes from a good understanding of the compound properties as they relate to multiple disease states through *in vitro* and *in vivo* screening, the ability to see the market application for what was initially thought to be an off-target effect, and for marketed compounds an in-depth understanding of off-label uses.[7] While an *a priori* unique insight to a potential new use is a desired starting point, this is not an approach that can be systematically applied for most FCCs.

It is likely that the most fruitful approach to explore repositioning opportunities for FCCs is through the systematic search for previously unappreciated properties, regardless of the reason for the original clinical failure. This can be done through identification of a new use based on mechanism of action or through a broad recharacterization of FCCs using both modern and classical tools. The latter is an empirical approach involving panels of tests designed to highlight previously unrecognized biological activities. Common approaches utilize panels of *in vitro* mechanistic assays (receptor binding, enzyme inhibition, transporter inhibition, etc),[8] *in vitro* functional assays (receptor function, cellular function, pathway perturbation, transcription profiling etc),[9] and/or *in vivo* animal models.[10] In most cases, these methods were developed originally for lead identification and characterization, but they can be readily adapted for expanded characterization of FCCs.

It is possible that a single test result pointing to a possible new therapeutic use will emerge from such broad recharacterization, although the more common result is a large and complex data set pointing to no obvious new therapeutic use. This is where bioinformatics and computational biology become essential tools.

Positioning an FCC within the space defined by therapeutic drugs using *in vitro* data

Existing drugs, taken collectively, occupy a unique landscape of therapeutic utility. Initial work by computational chemists made progress defining this space by using purely chemical descriptors.[11] However, a functional description of this landscape developed by using a combination of structure-based and biology-based descriptors appears to be much more useful,[7,12–15] especially in light of the recent observations that there are a limited number of successful molecular targets for approved drug substances, recently calculated to be 324 drug targets for small "drug-like" molecules.[16]

Most of the published validation of biology-based descriptors involves the use of *in vitro* mechanistic assays and is applied to *de novo* drug discovery and prediction of adverse effects;[7,12–14] however, the principles are readily applicable for searching for new indications for FCCs. The principle in this type of analysis is to allow biology, rather than chemical structure, to define similarity. Using biological descriptors, an FCC can be placed in the context of marketed drugs and reference compounds (i.e., based on biological effects, what are the most similar drugs to an FCC). This requires a broad and homogenous testing strategy covering drugs, reference compounds, and FCCs.

Identification of new properties using *in vivo* models

Alternative strategies for drug indication discovery and/or repositioning of FCCs take advantage of the inherent complexity of intact biological systems.[17] There are several recent discoveries of drug effects that would not have been predicted based on *in vitro* data alone. For example, the widely used statin class of HMG-CoA reductase inhibitors (e.g., atorvastatin and simvastatin) that is indicated for decreasing blood cholesterol levels also produces neurogenic stimulation in the hippocampus,[18] and the retinoic acid derivative, isotretinoin, inhibits microglial proliferation and neuroinflammation.[19] These potential alternative indications were uncovered by the use of an approach wherein stable isotopes are introduced into a whole animal or human model, and mass spectroscopy was used to measure components of various pathways.

Another approach that has been developed recently involves the use of a platform that consists of a variety of "gold standard" mouse models combined in a multiplexed fashion (theraTRACE® platform, Melior Discovery) to test for pharmacological activity that may have been overlooked previously.[20] A good example of the power of this approach is the discovery of MLR-1023 that is currently in Phase II studies for the treatment of type 2 diabetes.[5] This compound was created by Pfizer in the early 1970s, recognized as a cGMP-specific phosphodiesterase inhibitor,[21] and was tested in clinical trials for efficacy in the treatment of gastric ulcers. The compound was safe in clinical trials, but was not effective in decreasing time to healing compared to placebo and further development was discontinued in 1982. In 2005 the compound was tested in the theraTRACE® platform and was found to be very effective in decreasing blood glucose in an oral glucose tolerance test. There was no reason to expect that a cGMP-specific phosphodiesterase inhibitor would have such an effect. Subsequent testing in animal models of type 2 diabetes (db/db mice, Zucker rats) showed significant activity after single and multiple dose administration, and additive effects with metformin in lowering blood glucose levels.[22] The compound also blocked weight gain in mice on a high-fat diet and in mice treated for 28 days with the PPAR-γ agonist rosiglitazone. On the basis of these data, the compound was put into preclinical development by Melior Discovery and was approved to enter Phase II clinical trials. MLR-1023 is actually a novel activator of lyn kinase, a mechanism previously undetected for this molecule and discovered by screening against a panel of human kinases *in vitro*.[22]

Challenges in development of repositioned drugs

Existing data from an FCC can provide a significant advantage in the development. However, it is a mistake to assume that data used to support the original investigational new drug (IND) can fully support a new IND for the proposed new use. For example, the original indication for MLR-1023 was treatment of gastric ulcer, and the initial trials were conducted in the 1970s and early 1980s. Regulatory guidelines for the conduct of preclinical toxicology studies now require adherence to GLP standards that were not defined when the original toxicology studies were conducted, and clinical trial GCP requirements are more stringent than standards in existence in the 1970s and 1980s, necessitating new toxicology and Phase I clinical trials. In addition, analytical and bioanalytical methodologies and regulatory requirements have evolved significantly since the original manufacture of MLR-1023, requiring the development of new analytical and bioanalytical methods for the production of new compound and determination of stability. Reporting and drug approval requirements to whichever regulatory agencies are involved also need to be considered, and will factor in to decisions on regulatory strategy. Due to changes in regulatory standards, little data from the original Pfizer development effort in gastric ulcers could be used. However, the fact that the original data showed promising safety and pharmacokinetic profiles provides a level of confidence that is not available in NCE development.

In addition to starting development with significant biological data, repositioning FCCs can generally take advantage of past manufacturing and formulation development. This may greatly simplify early cost of goods and commercial feasibility assessments. However, careful consideration needs to be given to chemical intermediates for synthesis and excipients for formulation, as changes in technology and information about safety of the compounds involved may require revision of the synthesis or formulation to be used for the new indication.

Another key challenge in FCC repositioning is the protection of intellectual property. In the case of compounds where composition of matter patents exist, additional patents around method of use for the new indication may be possible, as well as patents around manufacturing and/or formulation if applicable. In cases where composition of matter patents have expired but the FCC has not been approved by regulatory agencies or marketed, such as MLR-1023, method of use patents provide strong protection because the compound can not be sold for the new indication by anyone but the patent holder or licensees. Sale of the FCC for another indication would require discovery of yet another new use and the supporting clinical trials, making off-label or generic competition unlikely during the use patent term.

Conclusion

The large number of failed clinical candidates (FCCs) presents a significant opportunity for potential repositioning of these molecules into new therapeutic areas for drug development. Screening methods to uncover new uses for FCCs include computational chemistry approaches based on similarities of structure, informatics to explore large sets of *in vitro* data on mechanism, pathway-driven exploration based upon metabolic changes caused by the FCC, and *in vivo*. Repositioning of FCCs represents a potential lucrative area of drug discovery that may allow the use of previously acquired data on the compound to accelerate the drug development process.

References

1. Sosei announces progression of SD118 towards phase 1 studies. February 21, 2007. www.sosei.com/en/news/pdf/PR_20070221-e.pdf
2. Ore Pharmaceuticals focuses resources on development of repositioned compounds. June 16, 2008. www.orepharma.com/ir-press-releases
3. Dynogen presents results of its positive phase 2a IBS-c Study with DDP733 (Pumosetrag). 22 May 2008. www.medicalnewstoday.com/articles/108338.php and Dynogen acquires exclusive rights to pumosetrag. Dec 1, 2004 http://findarticles.com/p/articles/mi_m0DHC/is_12_16/ai_n6358241/
4. Ore Pharmaceuticals acquires repositioned clinical-stage drug candidate from Roche. July 31, 2008. www.orepharma.com/ir-press-releases
5. Melior Discovery announces IND approval for novel diabetes drug MLR1023. March 16, 2009. www.meliordiscovery.com/pdfs/IND%20approval%203-09pdf.pdf
6. Ashburn, T. T. and Thor, K. B.. Drug repositioning: Identifying and developing new uses for existing drugs. *Nature Reviews Drug Discovery*. 2004;3:673–683.
7. Mason, J., Migeon, J., Dupuis, P. and Otto-Bruc, A. Use of broad biological profiling as a relevant descriptor to describe and differentiate compounds: Structure-*in vitro* (pharmacology-ADME)-*in vivo* (safety) relationships. In *Antitargets: Prediction and Prevention of Drug Side Effects*, ed. R. J. Vaz and T. Klabund (pp. 23–52). Weinheim: Wiley-VCH Verlag, 2008.

8. Root, D. E., Flaherty, S. P., Kelley, B. P. and Stockwell, B. R. Biological mechanism profiling using an annotated compound library. *Chemistry & Biology*. 2003;10:881–892.
9. Turner, S. M. and Hellerstein, M. K. Emerging applications of kinetic biomarkers in preclinical and clinical drug development. *Current Opinion in Drug Discovery & Development*. 2005;8:115–126.
10. Nidhi, Glick, M., Davies, J. W. and Jenkins, J. Prediction of biological targets for compounds using multiple-category Bayesian models trained on chemogenomics databases. *Journal of Chemical Information Modeling*. 2006;46:1124–1133.
11. Horvath, D. and Jeandenans, C. Neighborhood behavior of *in silico* structural spaces with respect to *in vitro* activity spaces – A novel understanding of the molecular similarity principle in the context of multiple receptor binding profiles. *Journal of Chemical Information and Computing Sciences*. 2003;43:680–690.
12. Fliri, A. F., Loging, W. T., Thadeio, P. F. and Volkmann, R. A. Biological spectra analysis: Linking biological activity profiles to molecular structure. *Proceedings of the National Academy of Sciences*, USA. 2005;102:261–266.
13. Fliri, A. F., Loging, W. T., Thadeio, P. F. and Volkmann, R. A. Biological spectra analysis: Model Proteome characterizations for linking molecular structure and biological response. *Journal of Medicinal Chemistry*. 2005;48:6918–6925.
14. Fliri, A. F., Loging, W. T., Thadeio, P. F. and Volkmann, R. A. Analysis of drug induced effect patterns to link structure and side effects of medicines. *Nature Chemistry and Biology*. 2005;1:389–397.
15. Overington, J. P., Al-Lazikani, B. and Hopkins, A. L. How many drug targets are there? *Nature Reviews Drug Discovery*. 2006;5:993–996.
16. Noble, D. From genes to whole organs: Connecting biochemistry to physiology. *Novartis Foundation Symposia*. 2001;239:111–123.
17. Shankara, M., King, C., Lee, J., *et al.* Discovery of novel hippocampal neurogenic agents by using an in vivo stable isotope labeling technique. *Journal of Pharmacology and Experimental Therapeutics*. 2006;319:1172–1181.
18. Shankaran, M., Marino, M. E., Busch, R., *et al.* Measurement of brain microglial proliferation rates *in vivo* in response to neuroinflammatory stimuli: application to drug discovery. *Journal of Neuroscience Research*. 2007;85:2374–2384.
19. Dutton, G. Repositioning idle drugs via systematic serendipity. *GEN*. 2008, January 1.
20. United States Patent 3,922,345 Lipinski, *et al.* November 25, 1975.
21. Ochman, A. R., Lipinski, C. A., Handler, J. A., Reaume, A. G. and Saporito, M. S. The lyn kinase activator MLR-1023 is a novel insulin receptor potentiator that elicits a rapid-onset and durable improvement in glucose homeostasis in animal models of Type-2 diabetes. *Journal of Pharmacology and Experimental Therapy*. 2013;342:23–32.
22. Saporito, M. S., Ochman, A. R., Lipinski, C. A., Handler, J. A. and Reaume, A. G. MLR-1023 is a potent and selective allosteric activator of lyn kinase *in vitro* that improves glucose tolerance *in vivo*. *Journal of Pharmacology and Experimental Therapy*. 2013;342:15–22.

Appendix I Additional knowledge-based analysis approaches

Raul Rodriguez-Esteban

Text mining tools

Tool evaluation

Before adopting any text mining tool, it is important to fully understand how they perform, especially in comparison to other tools available that have the same function. In order to benchmark text mining tools, however, one has to understand the performance metrics that are used to evaluate them and how these metrics apply in practical settings. The most common performance metric is the F-measure, which is the harmonic mean of precision (P) and recall (R):

$$F = \frac{2PR}{P+R}$$

Precision is based on the number of correct hits and the number of noisy hits that a tool produces. A tool with high precision minimizes the production of noisy hits (false positives, FP) with respect to the number of correct hits (true positives, TP).

$$P = \frac{TP}{TP+FP}$$

Recall is based on the number of correct hits retrieved versus those missed. A tool with high recall misses very few correct hits (false negatives, FN).

$$R = \frac{TP}{TP+FN}$$

Because the F-measure is the harmonic mean of precision and recall, a tool with high precision may still have low F-measure if it has low recall, and vice versa.

The F-measure was popularized at the fourth edition of the Message Understanding Conference (MUC), a competition organized to improve the state of the art in information retrieval (Chinchor, 1992). It was chosen for two reasons. First, because it encapsulates precision and recall in one measure, and second, because the harmonic mean favors tools that have similar precision and recall. The

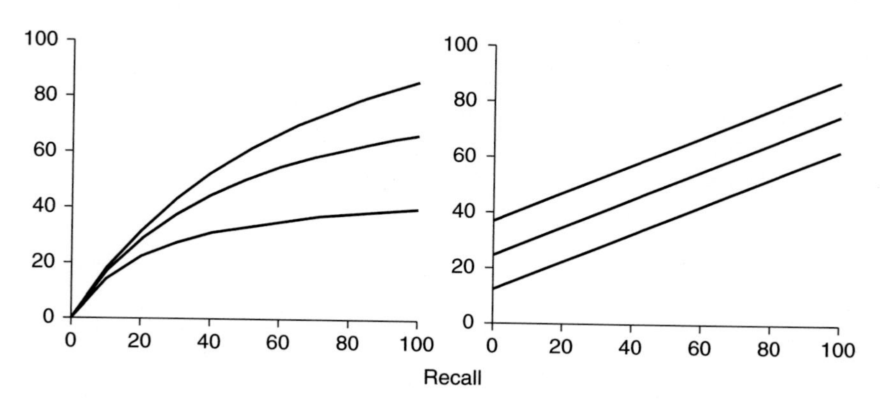

Figure A.1 Left, harmonic mean of precision and recall (F-measure). Right, arithmetic mean of precision and recall. Lines are plotted for three values of precision (25%, 50%, 75%).

harmonic mean discourages tools with either low precision or low recall (see Figure A.1). For example, a tool with precision 100% and recall 10% has an F-measure of around 18%, while the arithmetic mean of P and R is 55%. In this case, the harmonic mean penalizes the fact that P is much larger than R.

When approaching pharmaceutical questions, the F-measure needs to be taken cautiously for several reasons.

(a) The F-measure changes according to the "density" of facts of interest. When testing a text mining tool, the resulting F-measure will be typically lower when applied to a corpus with few facts than when applied to a corpus with many facts. For example, mining for genes in mechanical engineering patents is bound to produce much noise and few relevant results, regardless of the quality of the mining tool used. The area under the curve (AUC) is a similar measure of performance to the F-measure with the advantage that it does not have that drawback.

(b) Depending on the task, either precision or recall will be more important. The F-measure is a particular case of the F_β-measure in which $\beta = 1$.

$$F_\beta = \frac{(\beta^2+1)PR}{\beta^2 P + R}$$

Values of $\beta < 1$ are for cases in which precision is favored and values of $\beta > 1$ for cases in which recall is favored. For $\beta = 1$, precision and recall are equally valued. Depending on the amount of curation time available the value of β preferred will be different. If curation time available is ample, $\beta > 1$ can be a better choice.

(c) Sometimes it is most relevant to find the maximum number of *unique* facts because repeated results are irrelevant (Gomes *et al.*, 2006). Precision and recall can thus be redefined on the basis of unique TPs and unique FNs.

(d) The F-measure does not correlate linearly with the curation effort. The curation effort is related to the number of noisy (or spurious) results, which is also called overgeneration. A better measure of the burden of curation is one that I call "overhead," and which is closely related to precision. This measure is based on the number of spurious hits (overgeneration) divided by the number of correct hits:

$$\text{overhead} = \frac{FP}{TP}$$

For example, let's assume tools at break-even point, which means that they have precision equal to recall ($P = R$). One such tool has F-measure of 50%, and thus recall of 50% and precision of 50%. For this tool, the overhead is 100%. That means that for every correct result the tool produces, one has to also analyze and discard one flawed result.

A tool with recall of 75% and precision of 75%, however, creates an overhead of just 33%. On the other hand, a tool with recall of 25% and precision of 25% brings a 300% overhead (see Figure A.2). Thus, the difference between the curation effort associated to a tool with F-measure of 50% (overhead 100%) and a tool with F-measure of 65% (overhead 53%) is wider than it would seem at first sight from the difference in F-measures. Another implication is that F-measures above 75% (overhead < 33%) can be technically hard-fought but also without much consequence for curation, because the overhead introduced by the text mining tool is small. Very high F-measures (> 75%) can be useful in contexts in which there is no curation involved, but not otherwise.

Overhead is, in fact, a measure of the productivity of curation time. If curating a single text mining result takes a fixed time, then an overhead of 100% means that curating a correct result takes twice as much, because one needs to curate both a correct and a false result. However, it is often the case that curating a false result takes less time than curating a correct one. In such cases, overhead can be defined as:

$$\text{overhead} = \alpha \frac{FP}{TP}$$

where α is the ratio between the time it takes to curate a FP and the time it takes to curate a TP.

(e) The ranking of results usually matters. Psychologically, it is better for the individuals who analyze results, who might not be professional curators, to see the more relevant and precise ones first so that they can quickly appreciate their value and be more eager to go through them (Furrer *et al.*, 2014). This may also reduce curation time because the first results found may be enough for the purpose sought.

More broadly speaking, there is an important difference between *relevance* and precision/recall. A relevant result is a result that may have a number of certain desirable properties, for example, high-quality experimental settings, strong statistical proof, novelty or high validation. If a result has been very precisely identified in the literature but has limited relevance, then that should lower its ranking. On the other hand, a result that has been identified with low confidence but which could have high value (e.g., belonging to a journal or field with a high impact; Cokol and Rodriguez-Esteban, 2008) should be ranked higher. Thus, ideally, the results of a text mining exercise should be ranked according to multi-dimensional considerations. To aid in this process a number of measures exist to evaluate the quality of a ranking (Miwa *et al.*, 2013a).

Ontologies

Ontologies have been defined in many different ways, often depending on the scientific discipline in which they are used (Rzhetsky and Evans, 2011). For the purpose of text mining, ontologies are collections of linguistic elements such as concepts and names that are organized in ways that are useful for text mining. This definition puts the focus squarely on the usefulness of the ontology rather than on its particular structure and properties. An ontology for text mining can, in fact, take many forms. It can be as simple as a "flat" list of names, verbs, keywords, and expressions that has been created *ad hoc* for a text mining task. Such flat lists are also called vocabularies. Many of the vocabularies used in text mining are specific to a technical domain; for example, a list of laboratory devices. Such technical vocabularies are called terminologies and the elements that form a terminology are called terms. Terms will be discussed throughout this text because they are central for text mining. However, it is good to keep in mind that terms are often just names of things.

A text mining ontology can be made more useful by defining relationships between its members. These relationships are usually based on the knowledge we have about them. Thus, these relationships are *semantic*. Many ontologies used for text mining describe only a few types of relationships, such as the relationship "is a," as in "lymphoma is a disease"; or the relationship "has a," as in "a human has a cancer" (Rzhetsky *et al.*, 2000). Usually the relationships that an ontology admits are restricted by the creators of the ontology.

For example, the piece of knowledge that "lymphoma is a disease" can be used to build a simple ontology with three elements: the names "lymphoma" and "disease" and the relationship "is a" connecting them. This simple ontology could be easily expanded to cover the piece of knowledge that "psoriasis is a disease" by adding the name psoriasis to the ontology and reusing the relationship "is a" (see Figure A.3). With the addition of more disease names, this ontology could eventually become a comprehensive disease ontology.

This disease ontology has a simple structure. For many applications it would be more useful to have a more detailed structure that groups diseases according to their properties. For example, for a cancer researcher it would be convenient to have all the types of lymphomas in one place. Small categories can then be grouped to form larger categories following a hierarchy or "tree" structure. For example, B-cell lymphomas are a type of lymphoma, and lymphomas are a type of hematological

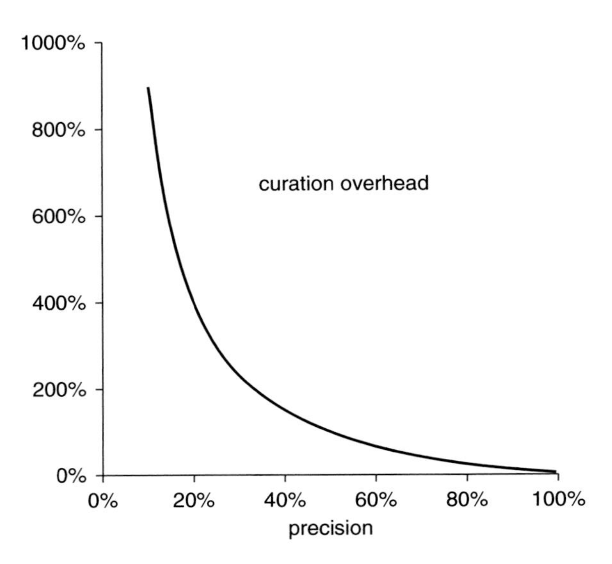

Figure A.2 Curation overhead depends non-linearly on the precision of a tool.

cancer. Thus, we can define the category hematological cancers as hierarchically above the category lymphomas, which is hierarchically above B-cell lymphomas. Ontologies with such hierarchical structure are called taxonomies, because they resemble the taxonomies traditionally used to classify species.

An issue with taxonomies is that there are often several useful ways in which their elements can be grouped. For example, lymphoma can be classified by its pathology as a cancer or by its location as an immune system disease. To address this, multiple taxonomical classifications can be assembled to create a polyhierarchy (or multiple hierarchy). A polyhierarchy allows terms to belong to multiple categories. Every hierarchy in a polyhierarchy brings its own "view of the world" to the table, grouping terms according to some property. Lymphoma can then be both a cancer and an immune system disease (see Figure A.3). Even existing taxonomies can be further enriched with the application of additional hierarchies. For example, several classifications of protein families, such as the PANTHER Classification System, can all be used to organize the NCBI Gene terminology.

A distinguishing feature of a text mining ontology is the availability of synonyms. Synonymy is a type of relationship that connects terms that refer to the same "concept." For example, the terms measles and rubeola can be connected by a relationship of synonymy because they refer to the same concept, in this case a disease. Synonymy, however, is not such a straightforward notion. Some synonymous terms refer exactly to the same concept and are interchangeable, such as the terms human and Homo sapiens. These "true synonyms" (or aliases) are rare in everyday language but fairly common in biomedicine due to an abundance of official names created by nomenclature committees. Terms which are near-synonymous, on the other hand, have slightly different meaning and might refer to more specific (hyponym)

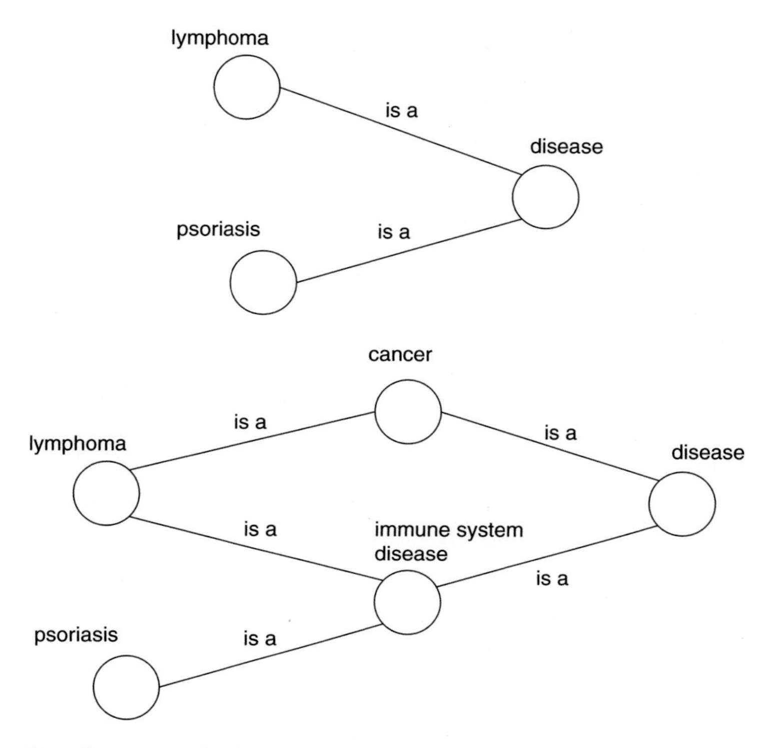

Figure A.3 Two disease ontologies.

or less specific (hypernym) concepts. For example, the name of a gene splice variant might be considered a synonym (hyponym) of a gene name in some contexts and not in others. Another special type of synonymy concerns spelling, dialectal and foreign language variants, such as the American English tumor versus the British English tumour.

Representing every nuance of synonymy in an ontology might be impossible (Hirst, 1995). To simplify, text mining ontologies typically consider synonymy as an all-or-nothing relationship, which is the practice used in many conventional thesauri. Usually, in a group of synonymous terms, one is chosen as the primary term to represent the group. Many text mining ontologies are in fact thesauri. NCBI Gene, for example, contains a thesaurus of official gene names and synonyms. The official gene symbol can be considered a "primary term." For example, in the case of gene *p53*, NCBI Gene lists the gene symbol *TP53* and several synonyms such as *TRP53*, *p53* and *LFS1*. Term normalization, which will be discussed in detail later, is about mapping a term found in free text to a concept or primary term from an ontology. Primary terms are sometimes equated with concepts. This can be useful, but it is worth remembering that a primary term might not be exactly equivalent to its synonyms due to the above-mentioned complexities of synonymy. Therefore, a primary term might actually represent a collection of concepts.

The breadth of an ontology is another important factor. As a rule of thumb, comprehensiveness is good: the more concepts and terms a terminology covers, the better for text mining. However, in large ontologies, concepts need to be organized in a useful manner so that they can be navigated and managed efficiently. The granularity of the categories cannot be too coarse or too fine for the purpose at hand. Every application will have a different need regarding nuance and diversity of terms.

A terminology that is comprehensive, enriched in synonyms and organized with a polyhierarchy can be a powerful ontology for text mining. Some of the most popular biomedical ontologies such as the National Cancer Institute (NCI) Thesaurus and MeSH ontology are organized in such fashion. The MeSH ontology and the NCI Thesaurus each organize terms in different ways and include different synonyms. For example, the NCI Thesaurus includes the term “rubeola infection” as a synonym for measles and rubeola, and designates “rubeola infection” as the primary term for the concept, while MeSH chooses instead “measles” as the primary term.

Biomedical ontologies that are not suited for text mining have few synonyms and contain many terms that rarely appear in free text. Often, these ontologies are primarily designed for annotation or for knowledge management. An example of such an ontology is the International Statistical Classification of Diseases and Related Health Problems (ICD), which does not include synonyms. Some ontologies contain many so-called “inverted terms,” which rarely appear in free text. An example of an inverted term is the term “Carcinoma, Neuroendocrine” from the MeSH ontology, which is far less likely to be found in free text than “neuroendocrine carcinoma.”

Despite the limited application that text mining has for ontologies with more complex structures than those so far described, it is worth getting acquainted with their potential. For example, some ontologies such as MeSH and SNOMED CT allow the composition of two or more terms to create terms that are not present in the ontology. MeSH, in particular, provides “qualifier terms” that can be used to modify MeSH terms. For example, an abstract that deals with the metabolism and pathology of lymph nodes can be annotated with the MeSH term “Lymph Nodes” and modified with the qualifier terms “metabolism” and “pathology,” leading to the composite annotations “Lymph Nodes/metabolism” and “Lymph Nodes/pathology.” This can be encoded in Medline XML like this:

```
<DescriptorName MajorTopicYN="N">Lymph Nodes</DescriptorName>
<QualifierName MajorTopicYN="Y">metabolism</QualifierName>
<QualifierName MajorTopicYN="N">pathology</QualifierName>
```

The MajorTopicYN parameter in the above XML is another feature of MeSH used to designate the importance of an annotation. In this case, the “Lymph Nodes/metabolism” annotation is a major topic of the abstract, while the “Lymph Nodes/pathology” annotation is a minor topic.

Managing ontologies

Ontologies need to be managed to keep track of changes in structure and new and deprecated terms. A number of ontology management applications exist such as OBO-Edit and Protégé. These applications can read and store ontologies using different ontology description languages. Fully featured ontology description languages such as the Ontology Web Language (OWL) are able to describe complex ontologies. However, because text mining ontologies have a simple structure, complex ontology description languages introduce unnecessary overhead. Specialized ontology description languages are simpler, have the necessary functionality and are easier to handle. Some examples are the Simple Knowledge Organization System (SKOS), the Open Biological and Biomedical Ontologies (OBO) format, and the ANSI/NISO Z39.19 format (often just referred to as the ANSI format). OBO is a popular ontology description language for biomedical ontologies. A compilation of biomedical ontologies in OBO format can be explored at the OBO Foundry.

Creating a useful text mining ontology automatically from scratch has so far proven to be unrealistic. Even enriching an ontology with synonyms is a very difficult task for computers to do automatically (Cohen *et al.*, 2005), although automatic term recognition (ATR) algorithms can help. Thus, ontologies need to be created with considerable manual effort. This effort, moreover, has to be sustained over time as the ontologies need to be updated regularly. The Unified Medical Language System (UMLS), which congregates many biomedical ontologies, gives a sense of the numbers involved. The UMLS groups equivalent and synonymous terms from different ontologies under single unique identifiers. Thus, the UMLS is a thesaurus that connects many thesauri, or a "metathesaurus." The UMLS unique identifiers are called Concept Unique Identifiers (CUIs). For example, the UMLS integrates the MeSH ontology and the NCI Thesaurus. The term measles from the MeSH ontology and the term rubeola from the NCI Thesaurus are both mapped in UMLS to the same CUI. The UMLS currently congregates 3 million such CUIs to integrate concepts from 168 biomedical ontologies covering around 9 million different names.

Identification of terms

Identification of terms is one of the main tasks in text mining. Krauthammer and Nenadic described term identification as a task that involves three steps which may be performed separately or simultaneously: term recognition, term classification and term mapping (Krauthammer and Nenadic, 2004). Term recognition involves recognizing all the terms that exist in a text. That means recognizing that a word or string of words designates something that could be of interest. Term classification involves assigning terms to a category or class, for example whether a term is a chemical or a disease. Term mapping, also called term normalization, involves mapping terms to a primary term (or concept) from an ontology, for example mapping a chemical name to a IUPAC identifier or a disease name to an ICD code. In practice, most term identification algorithms do not fit entirely within the recognition/classification/mapping framework, but it is a useful way to think about term identification.

Originally, the research on term identification in biomedical text owed much to the techniques developed for ATR (Kageura and Umino, 1996) and the MUCs (Grishman and Sundheim, 1996; Chinchor, 1998). One of the challenges presented at the MUCs involved recognizing several types of so-called "named entities" in text, which were defined as numerical expressions, temporal expressions and entity names. Entity names were proper nouns of people, organizations and geographic locations (e.g. John, United Nations, New York). Due to the success of the MUCs, named-entity recognition (NER) became a popular name for term identification, despite not being a very appropriate label. Thus, this chapter section could have been called "named-entity recognition" rather than "identification of terms."

Dictionary matching is the simplest approach to term identification. It involves finding all the mentions of a list of names from a dictionary that appear in a text. In text mining, this dictionary is typically a biomedical ontology, which can be enriched with synonyms. Obviously, the more synonyms the ontology contains the more mentions will be found in a text. However, abundance has its drawbacks. Very large ontologies add computational overhead for some applications. Some terms are ambiguous (polysemous) and can match useless terms (Hirschman *et al.*, 2002; Liu *et al.*, 2006). For example, NCBI Gene lists the gene *YES1* together with the synonym Yes. Ambiguity also results from terms in an ontology that have the same spelling (homographs). For example, the gene name *NIK* is a name for *MAP3K14* or a name for *MAP4K4*, depending on whether it represents the abbreviation for NF-kappaB-inducing kinase (*MAP3K14*) or for Nck-interacting kinase (*MAP4K4*). Another problem of dictionary matching is the existence of morphological variants of terms, such as the plural form, that may not be included in the ontology. This problem can be especially acute in languages such as Hungarian.

Certain techniques can be applied to improve the results of dictionary matching. In the case of genes, for example, matching can be improved by: (1) eliminating the most frequent ambiguous terms, such as terms which are frequent English words (such as Yes), and (2) performing fuzzy matching to capture morphological variations such as, in English, plural, hyphenated, spaced and capitalized forms (Rebholz-Schuhmann *et al.*, 2013). Fuzzy matching can be done with regular expressions, which are search patterns used in computer science. The use of regular expressions is popular in many text mining settings and regular expressions can even be developed automatically (Bui and Zeng-Treitler, 2014). Examples of dictionary-matching algorithms for biomedical text are MetaMap (Aronson, 2001), NCBO Annotator (Jonquet *et al.*, 2009) and ConceptMapper (Tanenblatt *et al.*, 2010). Some types of entities can be efficiently detected with dictionary matching such as cell types, cell lines, cell components and species, while others such as molecular function and biological process are not (Funk *et al.*, 2014).

The incompleteness of existing ontologies, the lack of observance by writers of naming conventions (Chen *et al.*, 2005) and the ambiguity of many names (for gene names, see Lovis *et al.*, 1995; Fukuda *et al.*, 1998; Cohen *et al.*, 2002) means that dictionary matching cannot identify properly all terms that appear in biomedical text. Alternatives to dictionary matching are statistical and rule-based algorithms, which can utilize different cues present in a text to recognize terms. For example, the word "gene" before the word "hedgehog" in the following sentence signals that hedgehog is the name of a gene and not the name of an animal:

Cloning, expression, and chromosomal location of SHH and IHH: two human homologues of the Drosophila segment polarity gene hedgehog.

A statistical algorithm can learn such cues, which cannot be recognized through dictionary matching alone. Dictionary matching, however, provides a baseline of performance that more advanced algorithms can improve upon. The ontologies themselves serve as an initial knowledge base for such algorithms.

Many algorithms for recognizing different types of terms have been developed, some free and some at a cost. Choosing one requires a careful assessment of the performance and the settings in which the algorithm was developed and evaluated. Statistical algorithms for term identification are typically developed with the help of a corpus that has been annotated manually. The algorithm learns using the examples from an annotated corpus, or "gold standard." A limitation of this approach is that performance decreases when the same algorithms are used outside of the corpus in which they were trained. This is an important problem given the diversity of biomedical language. Algorithms need to be developed using multiple training corpora to be robust (Rebholz-Schuhmann *et al.*, 2013), but creating corpora requires costly development. Many algorithms, instead, have been developed to compete in single-corpus challenges.

Another common disadvantage of statistical algorithms is their lack of transparency. Dictionary matching results are easy to understand by humans. Black-box or complicated statistical algorithms may produce puzzling results such as a term being identified in one place of a document and not recognized in another place of the same document. Moreover, there is a dearth of algorithms available for term identification beyond gene names, while there are ontologies of high quality for many types of biomedical terms ready to be used with dictionary matching.

Term identification algorithms may or may not perform term mapping/normalization. The advantage of normalization is that it allows easy summarization of results and integration with other data sources. A disadvantage of normalization is that it fails to recognize concepts not already present in an ontology and which may be novel. For example, a rare disease that has been recently discovered and does not appear in any ontology can be recognized but not normalized.

More importantly, normalization reduces the quality of final results (precision and recall) because it adds another layer of processing. A serial combination of text mining steps will usually result in lower overall performance because every step introduces errors. Thus, an optimal strategy to recognize terms could be to combine two term identification algorithms, one with normalization (which could be based on dictionary matching) and one without normalization that is able to capture novel terms.

Building a system for term identification in-house from scratch can be an arduous task, more so because terms in different disciplines exhibit different properties in the way they are created and in the frequencies they appear (Mihăilă *et al.*, 2012). Thus, using off-the-shelf solutions that require minimal modification can be of help. However, off-the-shelf solutions need to be evaluated in terms of their computational needs as some require large amounts of memory or processing power.

Genes and proteins

In text mining, genes and gene products such as proteins are usually considered a single class of terms due to the high naming ambiguity. A large percentage of gene names are also protein names and even humans cannot distinguish in many cases when such names are used as genes or as proteins (Hatzivassiloglou *et al.*, 2001). Here we will refer to gene names to imply both names of genes and names of their products. Much text mining work has been done to recognize gene names in text. A number of problems that make gene name recognition challenging have been described, such as defining where a gene name starts and ends (the name boundaries; Mani *et al.*, 2005) and the creativity with which gene names are coined (Seringhaus *et al.*, 2008).

NCBI Gene and UniProt Knowledgebase (UniProtKB) provide ontologies that can be used for dictionary-matching of gene names. Besides dictionary-matching, many statistical approaches have been devised for gene name identification. Fortunately, performance comparisons of some of those algorithms do exist and can be used to select one (Kabiljo *et al.*, 2009; Rebholz-Schuhmann *et al.*, 2013). An example of a high-performing open-source algorithm for gene name recognition is BANNER (Leaman and Gonzalez, 2008). For gene name normalization, evaluations across multiple corpora are not available and the choice needs to be based on the reported performance in competitions such as Biocreative III (Lu *et al.*, 2011). However, availability of participating software can be limited in many cases (see the discussion in Neves *et al.*, 2010). An open-source example of a popular gene name normalization algorithm is GNAT (Hakenberg *et al.*, 2011).

Chemicals and drugs

Several ontologies specialized in chemicals and drugs are available from IUPAC, RxNorm, the National Drug File, PharmGKB, the Anatomical and Therapeutic Chemical classification system, DrugBank, ChemIDplus from NLM, and ChEBI from the European Bioinformatics Institute (EBI). Many pharmaceutical companies have also developed internal ontologies for their own drugs, especially for the purpose of safety monitoring. Given the flexibility and variability of chemical names, statistical and rule-based algorithms are essential for their detection. Recently, there has been a surge in the area of chemical name identification, as shown by the chemical compound and drug name recognition (CHEMDNER) challenge from BioCreative IV (Krallinger *et al.*, 2013). Two freely available algorithms for chemical name recognition are CHEMSPOT (Rocktäschel *et al.*, 2012) and OSCAR (Jessop *et al.*, 2011). Because chemical name recognition and normalization have an important commercial application in areas such as patent analysis, a number of advanced algorithms also exist from companies such as ChemAxon and NextMove, both of which competed in CHEMDNER. Chemical name recognition can be performed with high quality, but normalization remains a major challenge. In the future it would be desirable to have a CHEMDNER-type competition involving chemical name normalization, which could also help benchmark existing commercial chemical name normalization tools.

Diseases

Identifying disease names in text is an easier task than identifying chemicals or genes due to the smaller vocabulary of disease names that exists. However, a number of issues still arise that make simple dictionary-matching suboptimal for the task (Doğan *et al.*, 2014). One is the usage of atypical or ambiguous disease acronyms. For example, the most common acronym for multiple sclerosis is MS, which is an ambiguous abbreviation. Other acronyms used for multiple sclerosis may refer to patient subpopulations, such as RRMS for relapsing/remitting multiple sclerosis, or be created *ad hoc* by authors, such as DMS for definite multiple sclerosis. Good algorithms and ontologies for identifying disease acronyms and name variations are valuable due to the importance of identifying disease names for pharmaceutical research. Besides ontologies such as the NCI Thesaurus, MedDRA and MeSH, algorithms have been developed for disease recognition (Jimeno *et al.*, 2008) and normalization such as DNorm (Leaman *et al.*, 2013).

Species

Some entities are of special interest because they offer contextual cues that modify the value of other entities. This contextual role depends on the question being posed. One such type of entity is that of species. For example, many homologous genes share names across several species. The gene symbol *TP53* is shared by human, dog, pig, chicken, and more. Information about homologous genes might be relevant to a different degree depending on the question. Besides humans, a toxicity question may prioritize results in rats and dogs, which are typical species in toxicology studies. Clinical studies may prioritize monkeys. Thus, the species used in an experiment is an important contextual piece of information. Species ontologies can be found in the NCI Thesaurus, MeSH, NCBI Taxonomy and others (Midford *et al.*, 2013).

Examples of species recognition algorithms are LINNAEUS (Gerner *et al.*, 2010) and SPECIES (Pafilis *et al.*, 2013).

Anatomy

In biomedical ontologies, the anatomy category typically covers not only body parts and organs but also terms for body regions, tissues, cell types, cell parts and other macromolecular structures. Similarly to species, anatomical locations can be important contextual cues that modify the value of other entities or facts. For example, the anatomic location for a disease (e.g. cancer of the lung versus cancer of the liver) or the cell types in which a protein is overexpressed. While anatomy vocabularies may be often quite static, some can be quite dynamic, such as those concerning cell types due to the constant discovery of new cell subpopulations and their markers. Ontologies such as MeSH, NCI Thesaurus and SNOMED CT have extensive anatomical vocabularies that can be used for dictionary-matching. Recently there has been renewed interest in algorithms to identify anatomical terms (Miwa *et al.*, 2013b; Xu *et al.*, 2014).

Experimental methods

Experimental methods are useful to qualify the value of an experimental result. For example, some experimental methods might be deemed less reliable than others and thus produce data that are less trusted. Moreover, experimental scientists are interested in knowing the particular methods used by other experimentalists to produce a result. In scientific publications, the methods section is especially rich in description of such information. Explanations of experimental methods may cover a wide range of aspects, such as the description of procedures, cell cultures, breeding of animals, and purchase of reagents from vendors. Some of the terms used in the description of methods concern devices, assays, cell lines, animal models, compound leads and anatomical names.

Ontologies for terms used in experimental methods are more fragmented and less well-defined than for other terms. Moreover, some of the terms are company- or laboratory-specific, such as device names and compound lead names. Many pharmaceutical companies have their own lead naming system and may use these names in publications. Cell lines are a bright spot due to the availability of ontologies for dictionary matching such as Cell Line Knowledge Base, Cell Name Index, and Cellosaurus. Cell line names tend to be distinctive and have less variation and thus are easier to identify than other types of names. Assays, on the other hand, are especially difficult to identify as they are not usually described by name but by a description of steps and procedures. Thus, classifying and normalizing assay names might be unrealistic.

There is also a dearth of ontologies for animal models. Fortunately, though, the majority of drug discovery studies focus on a reduced number of animal models, which can be covered with an *ad hoc* ontology. Identifying novel animal models is less of a priority as drug discovery is often more interested in established animal models than in those that have not been proven yet.

Biomarkers

A biomarker can be a wide range of things, such as temperature changes, chemicals in the urine, and blood metabolites. In pharmaceutical drug discovery, however, there is usually higher focus on clinical biomarkers that are secreted proteins in blood, serum, and urine or on mechanistic and functional biomarkers that are downstream genes of a therapy under study. These biomarkers are often specific to ongoing drug development projects, but are sometimes sought as alternatives to established biomarkers. Because biomarkers are defined in terms of their relationship to a drug or disease, they are better understood in the context of relationships, which are discussed later.

Phenotypes and adverse events

Phenotypes and adverse events, like biomarkers, are categories of terms that are easy to define yet difficult to list comprehensively. Human evaluators tend to disagree on what constitutes a phenotype (Chen and Friedman, 2004). The classification of an event as an adverse event, on the other hand, depends on the precise context. For example, an event might be considered adverse for a given drug and therapeutic for another drug. The main ontologies for adverse events are MedDRA, which is a reference ontology for safety monitoring, and SNOMED CT, which has a focus on pathology.

Phenotypes, the best defined of which are diseases, cover a wide range of terms and descriptions that can be species-specific. The focus of existing phenotype ontologies is more commonly the annotation and organization of phenotypic results from experimental studies, such as the Mammalian Phenotype Ontology (Smith *et al.*, 2005) used by the Jackson Laboratory for mouse and rat experiments, or the Human Phenotype Ontology (Köhler *et al.*, 2014).

Mutations

With the growth of personalized medicines that are tailored only to patients with certain mutations, mutations have been gaining in prominence in drug discovery. Such mutations range from single point mutations, to deletions, translocations, amplifications, etc. Mutations are often described in the text in non-standard ways but still follow certain patterns (see table 1 in Rebholz-Schuhmann *et al.*, 2004). These patterns can be largely captured using rules. A number of publications have shown that mutation identification algorithms can reach high performance. An example of mutation detection algorithm is tmVar (Wei *et al.*, 2013), which is available for download. A harder task is to associate mutations to the genes and phenotypes to which they are related, or to identify their prevalence in certain patient populations. This is more similar to finding relationships. One approach proposed for gene-mutation extraction is to use crowdsourcing to curate text mined results (Burger *et al.*, 2014).

Measurements and parameters

Measurements and parameter values are usually described in the literature with numbers and measurement units, such as 1 liter, 2 nM and 3 mg/kg. Such quantities can be identified in the literature largely with the aid of regular expressions and they tend to appear in the full text of articles rather than in the abstracts. Due to the costs of accessing full text, strategies to identify potentially relevant full-text articles using evidence from abstracts can be cost-effective (Hakenberg *et al.*, 2004). Of special interest in drug discovery are those measurements and parameters related to systems biology and pharmacokinetic modeling. Those values, however, need to be associated to the proteins, pathways or diseases that they describe. A molarity value might be easily recognized in a text, but it is harder to recognize that it describes, for example, the dissociation constant (Kd) of a protein in a certain disease. Linking such contextual information poses a formidable challenge for text mining.

Companies and other institutions

Identifying the names of companies and other institutions has practical value for many pharmaceutical applications, such as competitive intelligence, partnering and licensing. Company and other institution names can be often captured using ontologies and regular expressions. Company and institutional structures can be represented in ontologies with a hierarchical tree structure of subsidiary names, branches and units. Such ontologies need to cover name variations and abbreviations, as well as the outcome of acquisitions, mergers and name changes. For example, an ontology about the company F. Hoffmann-La Roche may include the names of acquired companies Genentech, Chugai and Boehringer Mannheim, as well as the names of the divisions Roche Pharmaceuticals and Roche Diagnostics. An ontology for the University of California system would include the name of its many campuses, such as University of California, Berkeley and its synonym UC Berkeley.

A helpful circumstance is that specialized fields for author and/or institutional affiliation exist in many documents, such as patents, clinical trials and scientific publications; which reduces false positives when identifying such names. Most Medline records since the 1990s have at least one author affiliation and more recently register multiple affiliations (as is done in Embase). Location information such as postal code, city and country can be used in conjunction with other textual clues to improve recognition and mapping. Moreover, many companies use common acronyms as part of their names, such as LLC: Ltd. and Co. in the USA; S.A. in Spain and France; and GmbH in Germany. Some institutions include revealing keywords in their names such as clinic, pharmaceuticals, hospital and foundation (or versions of these names in other languages). Thus, a combination of ontologies, regular expressions and other rules can be helpful towards identifying and mapping institutions (Jonnalagadda and Topham, 2010).

Authors

The main challenge when identifying author names is disambiguation due to the fact that many names are shared by multiple authors. Disambiguation can be performed at database-level (such as across the whole Medline) by clustering publications according to authorship features such as affiliation information, keywords and annotations. Because authors change affiliations and scientific fields over time, disambiguation can only be performed approximately, but the performance of disambiguation algorithms is still quite good (Torvik *et al.*, 2009; Varadharajalu *et al.*, 2011; Liu *et al.*, 2014). Only the widespread adoption of unique author IDs, such as the Open Researcher and Contributor ID (ORCID), could completely solve this problem.

Abbreviations

Mapping abbreviations (also known as "short forms," a type of which are acronyms) to their definitions ("long forms") is a well-understood problem that can be solved with high precision using rules (Wren and Garner, 2002; Pustejovsky *et al.*, 2001). The first step in the process consists of finding the location in the text in which an abbreviation is defined. For example, a document may use the acronym MS for multiple sclerosis and define it with the words: "multiple sclerosis (MS)." Resolving abbreviations can be done effectively and it is advisable, in fact, to expand every abbreviation in a text to its long form as a routine pre-processing step. This transformation improves the mining because definitions are more meaningful and less ambiguous than abbreviations, which can be polysemous or used only by one author. Abbreviations that do not include their definition within a document are more challenging to resolve and may require the use of abbreviation dictionaries (Chang *et al.*, 2002).

Relationships

The biomedical literature describes many relationships between pairs of entities that are of interest for drug discovery. A well-studied type of relationship is the protein–protein interaction (PPI), which in text mining has been loosely defined to include an ample number of relationships between proteins beyond physical interaction. Thus, PPIs also cover indirect relationships between proteins. Besides PPIs, other relationships of interest are, for example, those between drugs and proteins, drugs and drugs, drugs and phenotypes, protein expression and cell types, mutations and diseases.

Strategies to identify relationships start with term identification. Therefore, precise term identification is a prerequisite for precise extraction of relationships. Once terms have been identified, potential relationships between them can be investigated using statistical and rule-based approaches. The simplest statistical method to identify relationships is co-occurrence within a document. The rationale behind document co-occurrence is that two terms that appear in the same document are probably related to some extent. A corollary to this rationale is that if two terms co-occur in many documents then those terms are strongly related. The main problem with co-occurrence is that it produces many FPs for terms that co-occur in a document yet have minimal or no relationship. FPs can be reduced by tossing away statistically weak co-occurrences, for example involving terms that are very frequent (Hu *et al.*, 2003).

Document-level co-occurrence might seem an unsophisticated method to establish the relationship between two concepts; however, due to its simplicity and robustness it can have an important role in situations in which curation is not a constraint, and also for high-throughput applications. For example, in a high-throughput screening (HTS) a library of perhaps several million compounds can be prioritized for compounds that co-occur with certain phenotypic terms in the literature. In such a case it might not be necessary to determine whether there exist strong compound–phenotype links beyond co-occurrence.

A more targeted approach is to look for co-occurrences only at the sentence level or within a window of words. In this approach, only terms that appear in the same sentence, or separated by at most a certain number of words, are counted as co-occurring. This strategy reduces the number of FPs, but it also increases the number of FNs because some relationships do exist between terms in different sentences or separated by large word distances. Examples of relationships that exist across different sentences are those that involve co-references. Co-references are expressions or words such as pronouns that are used to refer to other terms. In this example the co-reference "it" refers to the term "hATF4":

In this report, we show that hATF4 is a strong activator in both mammalian cells and yeast. It interacts with several GTFs and with the coactivator CBP […] (Liang and Hai, 1997)

Relationships can also exist across sentences without the intervention of co-references because the reader is able to understand from the context, or what in NLP is called the *pragmatic* dimension of the text. For example, in an article dealing with inflammatory myofibroblastic tumor (IMT) which says:

Recent detailed studies [2,6] suggested that IMT is a neoplasm with benign or low-grade malignancy. The exact incidence of this disease is unclear. (Takeda *et al.*, 2008)

The reader understands from the context that "this disease" is a reference to IMT. Looking for relationships at the sentence level, however, might be the most effective approach for extracting interactions (Ding *et al.*, 2002). For that, the task of detecting sentences itself needs to be performed first. For that purpose there exist so-called sentence splitters which are specialized for biomedical text, such as the JULIE Lab Sentence Boundary Detector (Tomanek *et al.*, 2007).

To improve on the sentence co-occurrence method, further textual elements from the sentence can be considered. For example, the presence of certain keywords can increase the likelihood that a sentence describes a relationship between two terms. This idea is used by the tri-occurrence method, which consists of identifying sentences that harbor two terms plus a relevant keyword. A good example was shown by Zhang and colleagues, who built a detailed statistical approach to detect biomolecular interactions using tri-occurrence (Zhang *et al.*, 2013). The selection of keywords for tri-occurrence needs to be done carefully to balance recall and precision. Tri-occurrences (or, to generalize, *n-occurrences*) can also be considered at the document level, or a mix of sentence and document level. Keywords end up working much like filters, which will typically affect the precision/recall values (Milward *et al.*, 2006).

Yet another way to identify relationships involves analyzing syntactic and semantic elements at the sentence level. Such an approach typically starts by identifying co-occurrences within sentences. Once these are identified, additional sentence information is considered, such as whether the sentence includes verbs or verbal forms that are commonly used to describe relationships, which are sometimes called "anchor verbs" (Hatzivassiloglou and Weng, 2002). For example, there are verbs which are often used to describe PPIs, such as "interact" or "bind" (Sekimizu *et al.*, 1998, Rodriguez-Esteban *et al.*, 2006; for a list of verbs see Kabiljo *et al.*, 2009; Li *et al.*, 2014). These verbs can indicate the nature of an interaction, whether it is direct (physical) or indirect (logical), or whether it is activating or inhibiting. Additionally, verbs may connect the co-occurring terms syntactically. For example, in this sentence: *APC binds to the novel protein EB1.* (Su *et al.*, 1995), "APC" is the subject and "the novel protein EB1" is part of the predicate. The verb "binds" connects these two terms. This interaction can be represented with a triplet such as (APC, EB1, bind).

More elaborate strategies to identify relationships try to capture complex syntactic patterns and/or use machine learning algorithms. Syntactic patterns can be described using flexible descriptors called templates. Typically, many templates need to be built to enumerate the different ways in which relationships appear in text (Friedman *et al.*, 2001). Then, every sentence that matches one of the templates is said to contain a relationship. A very simple template such as *[protein name phrase] [interaction verb] [preposition] [protein name phrase]* would match the sentence "APC binds to the novel protein EB1." Each template is a "rule" and thus using templates is a rule-based approach. Templates can be quite complex and use multiple types of information to produce a match, such as part of speech information (e.g., adjective, noun, adverb) and information about dependencies between the elements in the sentence.

Templates can also be enriched with ontologies. For example, the template *[protein name phrase] [interaction verb] [preposition] [protein name phrase]* could be associated to an ontology of verbs for protein interactions and an ontology for protein names. Furthermore, templates can be refined with term co-occurrence statistics, which is the approach followed by the STRING database (Saric *et al.*, 2006; Franceschini *et al.*, 2013). An open source library for building templates is OpenDMAP (Hunter *et al.*, 2008).

The highest performing machine learning algorithms for identifying PPIs are kernel-based (Tikk *et al.*, 2013). Machine learning algorithms, like templates, use syntactic and semantic information from sentences, but they also need to be trained in manually curated corpora. As is the case with machine learning algorithms for term identification, these algorithms decrease in performance when they run in corpora different from the manually curated corpora in which they were originally trained (Pyysalo *et al.*, 2008). Even factors such as variations in the distribution of sentence lengths can have an impact on their performance (Tikk *et al.*, 2013).

The low F-measures produced in the PPI extraction challenges in BioCreative II (Krallinger *et al.*, 2008) indicate that in many cases using such algorithms might not be justified at the moment. This is due, in part, to the fact that these algorithms rely on term identification as an initial step and then add a further layer of processing, resulting in a decrease in overall performance. Machine learning algorithms to detect PPIs could benefit from focusing on more specialized types of interactions (Tikk *et al.*, 2013) or by training in larger data sets beyond the typical small hand-labeled corpora that can be affected by curator biases (Rinaldi *et al.*, 2014). An example of a more successful task than PPI extraction is the extraction of drug–drug interactions (Herrero-Zazo *et al.*, 2013).

Depending on the situation, it might be practical to use simple high-recall, low-precision approaches such as sentence tri-occurrence rather than approaches with higher recorded F-measure but which are more complex and less robust. Rule-based algorithms also have the advantage that they are not black-box and it is easier to tailor them to narrow problems, such as extracting phosphorylation

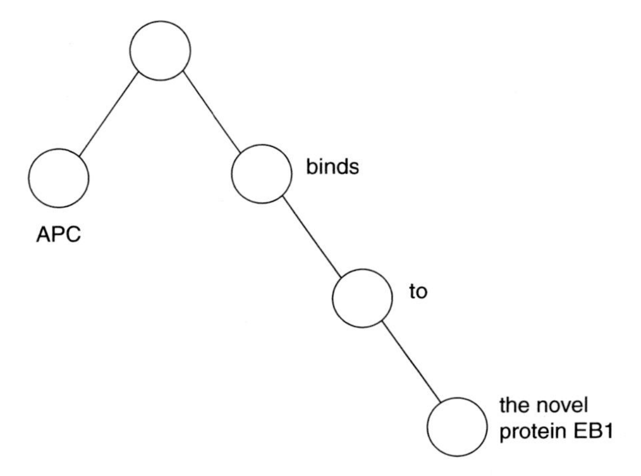

Figure A.4 Deep parsing structure of the sentence "APC binds to the novel protein EB1."

interactions (Hu *et al.*, 2005; Torii *et al.*, 2014). A commercial tool called I2E from Linguamatics (Milward *et al.*, 2005) relies on templates that use syntactic information and ontologies, although recently it has added some machine learning capability to identify gene names. I2E is popular in pharmaceutical companies and allows users to create and modify templates interactively using a graphical interface and to build *ad hoc* ontologies that are appropriate for a particular task, something that Linguamatics calls "agile text mining."

Co-occurrence, template and machine learning methods have in common that they all try to detect some measure of relationship. The results from two well-constructed analyses, one using sentence co-occurrence and another using sentence tri-occurrence, will not usually differ in overall shape except that tri-occurrence will yield lower recall and (hopefully) higher precision. Similarly, a co-occurrence within a window of words of smaller or larger size will not produce an altered overall picture. This can usually be expected regardless of the approach used as there exists a "relationship continuum." The strength of a relationship depends on many factors and it is not a binary proposition. In other words, relationships, like any other fact, are not absolute. Some relationships are just hypotheses or can be inferred from indirect evidence. For example, two proteins that belong to the same pathway might be assumed to be related. Other relationships are described in one article and then negated in another article. Yet other relationships are well established in numerous articles. In the end, the goal of text mining is to assess the strength, or better, the value of a relationship, especially from the point of view of the task at hand. Even speculative relationships might be interesting in some contexts due to their potential novelty (Malhotra *et al.*, 2013). Every approach discussed will yield some results and typically the strongest relationships will surface regardless of the approach chosen. The best approaches will allow the reduction of noisy hits and the recognition of infrequent facts.

Pathways

Relationships that share terms can be connected to form networks, such as PPI networks and even interactomes (Iossifov *et al.*, 2009). One type of network which is important for pharmaceutical research is the signaling pathway. This is due to the fact that much reasoning about the therapeutic effect of drugs on disease is pathway-centric. Signaling pathways are neatly described in textbook figures, pathway databases and review articles. However, pathway definitions vary greatly depending on the pathway curator. The presentation of pathways in text is also difficult to interpret as pathway information is mentioned erratically (Wu *et al.*, 2013). A way to produce something out of this situation is to integrate pathway information from text with pathway diagrams from databases to derive, for example, associations of diseases and pathways (Li and Agarwal, 2009). Unlike signaling pathways, metabolic pathways are better understood and can be extracted from the literature using approaches similar to those used for PPI extraction (Czarnecki *et al.*, 2012).

Causal relationships and perturbations

Causal relationships are a subtype of relationships (Mihăilă *et al.*, 2013). Discovering causal relationships in disease is particularly valuable, for example when looking for particular mutations that cause a disease. Causal relationships can also indicate mechanisms, pathways or signaling cascades that can be used as biomarkers. A cause can be a point of intervention, something that can be changed. In fact, causal relationships can also be artificially produced by humans, as in therapeutic and experimental interventions. Such causal relationships are called perturbations (Rodriguez-Esteban *et al.*, 2009). An example of a perturbation would be the modulation of gene expression leading to a change in an animal phenotype.

The description of the experimental method used in a perturbation can indicate the nature of the intervention in more detail. Examples of gene perturbation methods are siRNA inhibition, blocking antibodies and gene amplification. Many perturbations can be classified according to their directionality; that is, whether they increase or decrease the quantity or activity of the perturbed element, such as the activity of a protein or the concentration of a metabolite. Ascertaining the directionality is important because many drug discovery technologies are more suited to either activate/increase/agonize or to inhibit/reduce/antagonize. Mining perturbations opens a window into the different biological interventions that have been performed by researchers and especially those that could have therapeutic application.

Events

Relationships are too simple to describe many biological phenomena in detail. The concept of a biological *event* has been created as a catch-all concept that includes multiple types of biological phenomena mentioned in text, with special focus on molecular events (Kim *et al.*, 2008). Some examples of molecular events are post-translational modifications, expression of genes and epigenetic events. Event mining is also concerned with elements that qualify events, such as negation (polarity) and degree of certainty. Such elements have been called meta-knowledge (Thompson *et al.*, 2011) or a multidimensional way to classify text (Shatkay *et al.*, 2008).

Events can be described with representational languages that connect the different aspects that form an event. Examples of events and their representation can be found at the BioNLP Shared Tasks competitions for event mining. The practical application of event mining to pharmaceutical research is still limited. General event mining algorithms have performance limitations, perhaps due to the complexity of the task, yielding best F-measure values in the 50%–60% range (Pyysalo *et al.*, 2012). Event mining strategies may need to be tailored to specific events in order to produce useful results for pharmaceutical applications rather than trying to cover multiple types of events. An interesting resource is the literature-wide event database EVEX (Van Landeghem *et al.*, 2012).

Question answering

While still not in wide use in drug development, question-answering tools have raised their profile due to the success of IBM's Watson system. A tradition of question-answering tools, however, already existed in biomedicine before Watson's arrival, such as those tools developed for the TREC Genomics Track (Hersh *et al.*, 2006). Question-answering systems are able to (1) understand free text questions, and (2) retrieve the best answers to those questions from within a knowledge base or text corpus. These two aspects of question answering are of different value in the drug discovery setting. The first aspect (understanding questions) could allow users without prior training in bioinformatics, text mining or information science access to a wide range of resources. The second aspect (answering questions) has more limited applicability as text mining and other disciplines are already concerned with that. It is nonetheless possible that through combining many question–answer pairs, complex issues could be better analyzed to support certain decision-making processes (Yang *et al.*, 2014).

Question-answering systems could be fine-tuned for specific questions and offered to many more users that current text mining tools. Question-answering interfaces with limited options could narrow user choices in order to make them manageable. An example of how this could work is the interface used by Ingenuity Answers, which is a querying interface for the Ingenuity knowledge base. Ingenuity Answers offers dropdown menus that allow users to construct questions with clear constraints and a narrow number of choices.

The process of text mining

Know thy users, for they are not you

A person with some text mining expertise (for short, a text miner) may not be the final recipient of the analyses produced by text mining tools. Often the recipients are other experimental or computational biologists working alongside the text miner in a project. This collaborative environment means that it is crucial for text miners to, as the saying goes, "know thy users, for they are not you." A text miner should understand the needs and expectations of collaborators, and should channel those needs and expectations with regard to several aspects. Some of those aspects are as follows.

Noise

False positives (noise) are an inevitable outcome of text mining processes. Some collaborators have low tolerance for noise and should only be presented with high-quality results. An example of a high-precision approach can be seen in Xu and Wang's mining of drug–disease interactions (Xu and Wang, 2013). Other collaborators only desire digested and summarized content ready for presentation. Yet

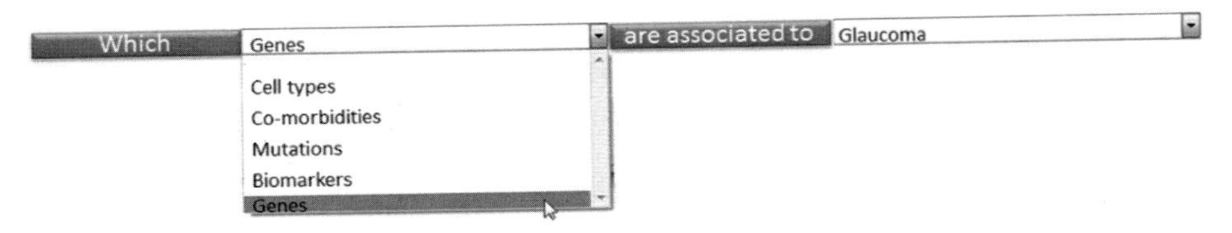

Figure A.5 Simple question-answering interface inspired by the Ingenuity Answers interface.

other collaborators are more hands-on and are willing to sift through reasonable amounts of false positives provided that the effort is not strenuous and the density of interesting results is high enough. The results presented to these hands-on collaborators may only need light curation. In general, it is advisable that noise in text mining results be lower than the noise tolerance of the person analyzing them.

Speculative and negative facts

Speculative and negative facts are less of a problem in pharmaceutical settings than is usually thought. Unlike other types of mining errors, collaborators understand the reasons behind them. Moreover, such facts can be informative as well. After all, any piece of evidence from a text mining analysis is just a starting point, not the end, for further study. A negative or speculative fact can lead to other insights. Negative results, for example, are rare and may concern important facts that have been falsified.

Recall

At times, some collaborators expect maximum recall. In those cases, no results can be missed. More typically, however, collaborators have different expectation levels which depend on the task at hand. Understanding and setting expectations of recall is an important step in the text mining process. Between the extreme requirements of "very high precision" and "very high recall," there is a range of choices.

Discerning the attitude and expectations of collaborators can only be done through an interactive process. Initial questions posed by collaborators are often incompletely defined and need to be steered and translated by the text miner into more precise terms that can be addressed with the aid of text mining approaches. In fact, many common questions can be answered by adapting proven strategies to the particulars of a case.

Text mining does not replace reading

While text mining results can be helpful to experts in any field, regardless of their depth of expertise, it does not replace actual reading. It is always important for text miners to familiarize themselves with the writing style and vocabulary used in documents that are relevant to a certain question. Such reading helps build, for example, syntactic templates, *ad hoc* ontologies and corpora. There can be systematic flaws such as ontologies lacking important synonyms and algorithms missing syntactic structures that human reading can quickly notice. Furthermore, the process of curating results can involve considerable reading, depending on the size of the results. Text miners need to read text before and after mining it.

Pre-processing

Pre-processing steps may seem routine, but they ultimately affect the quality of results. Unlike in most published text mining studies and competitions, everyday pharmaceutical text mining needs to deal with text from heterogeneous origins, variable shape and different format, such as PDF, HTML and XML documents. As mentioned, scientific articles in PDF format bring a host of problems and specialized tools have been developed for this purpose. PDF documents from other origins, such as conference abstracts, internal documents and patents, are problematic as well. HTML documents are less problematic, but need to be properly parsed. XML documents often lack some formatting information that could be useful for text mining, such as markups for superscript and subscript, font weight and size, equation definitions, footers, headers, table structures and hyperlinks. Moreover, XML schema definitions change over time and XML (and HTML) documents might be malformed.

Character encoding can also be a problem. Multiple encoding standards are prevalent, such as those from the UTF and ISO families. Characters can also be represented using HTML and XML character references. Within these character references, for example, the Spanish character ň might be represented as *ñ*, *ñ* or *ñ*. A single document may contain all three representations. Characters such as <, > and & can easily make an XML parser fail if they have not been escaped as *<*, *>*, and *&*, respectively. Character escaping has to be applied carefully as one can often see cases of double-escaping such as *&* or *<*. Managing documents in multiple languages elevates these character encoding problems to another level.

Patents represent a number of challenges on their own. Chemical formulas can be long-winded and trigger many false positives when not recognized as such. Claim dependencies and Markush structures add a complex layer of self-reference. Patent claims can harbor particularly long sentences; for example, a claim can mention scores of disease names that are supposed to be relevant for an invention. Moreover, interest has been growing to analyze patents in non-English languages, such as Chinese patents. For all these reasons and more, patents are one of the frontiers of text mining.

Defining scope

One of the most important steps in text mining is to set the scope of a text mining analysis. First, it is necessary to choose the corpora that will be explored, such as abstract, patent or clinical trial databases, depending on the question and the resources available. From those corpora, only a subset of documents might be explored in detail. For example, US patents are a very large repository of information covering many topics beyond biomedicine. Focusing only on patents that mention biomedical terms such as names of diseases provides a more manageable set of documents and may produce fewer false positives without much reduction in recall.

Such filtering can reduce the level of term ambiguity within the corpus. For example, the gene name wingless is less ambiguous when it appears in the same documents as its synonym wg or as the species name *Drosophila melanogaster*. Thus, choosing documents that mention wg leads to less ambiguity for the name wingless. Pairs or groups of terms are less ambiguous in their meaning than individual terms.

There are other ways in which the scope of a corpus can be delimited, such as focusing on documents from certain years or on specific sections of documents. The scope can also be restricted to documents manually annotated with certain terms, such as Medline abstracts annotated with certain MeSH terms or EMBASE abstracts annotated with certain EMTREE terms. A corpus of abstracts focused on type 2 diabetes could be gathered by selecting all Medline abstracts annotated with the MeSH term "Diabetes Mellitus, Type 2" in Medline. Such manual annotations can be used as effective filters.

Setting the scope increases in importance with the cost of gathering and processing documents, for example in the creation of full text corpora, which can be costly or time-consuming to build. One effective strategy to select articles for full-text download is to use clues in the abstract that indicate the likelihood that a full-text article contains relevant information. Statistical approaches have been used to classify abstracts on whether the full-text article is likely to contain, for example, kinetic parameters for systems biology (Hakenberg *et al.*, 2004) and PPIs, as seen in the Biocreative II (Krallinger *et al.*, 2008) and III (Krallinger *et al.*, 2011) competitions.

Interoperability

The proliferation of text mining tools that perform specific subtasks means that interoperability solutions can help decrease maintenance and pre-/post-processing steps in text mining pipelines. For such purpose there exist interoperability frameworks such as GATE and UIMA, and pipeline processing software such as the open-source Knime and the commercial Pipeline Pilot and Inforsense Suite, which also provide specialized text mining components. A drawback of these approaches is that they require a certain level of expertise and maintenance. UIMA is perhaps the simplest of them as it is, at its barest, a set of Java functions that standardize input/output formats. The full benefits of UIMA come with components that are completely UIMA-compliant or with the use of UIMA AS/DUCC for parallel processing. A simpler solution for interoperability involves using a common XML format tailored for text mining such as the BioC format (Comeau *et al.*, 2013).

Post-processing and curation

As Gomes and colleagues have pointed out: "One cannot expect text mining to produce accurate, final knowledge with no need for human review" (Gomes *et al.*, 2006). Curation is an integral part of the text mining process in drug discovery and should influence the choice of text mining strategies. Availability of curation time should be considered as one of the main factors in a text mining task.

Curation environments have been developed within pharmaceutical research environments that integrate text mining capabilities (Rinaldi *et al.*, 2008). One of the advantages of such integrated systems is that they can provide feedback on the quality of the text mining tools as well as suggestions for the improvement of, for example, the ontologies being used.

Because every processing step in text mining often decreases recall, approaches that are more "minimalist" might be favored at times. However, such minimalism increases the curation time so that a "drive till you qualify" (or "text mine until you can curate") compromise has to be found in which text mining steps and filtering constraints are applied until results become manageable for curation. This approach is related to the concept of "assisted curation" (Rinaldi *et al.*, 2014).

Because many text mining tools and strategies do not provide ranked results, part of the curation process involves organizing the results so that they can be prioritized for further analysis. Semi-automated (Jamieson *et al.*, 2013) or automated (Rodriguez-Esteban *et al.*, 2006) curation approaches could help in that process.

A different type of curation is that performed by biocurators, who are expert biological knowledge curators that work in the creation and maintenance of biological databases. There is a history of collaboration between the text mining community and the biocuration community (Hirschman *et al.*, 2012; Lu and Hirschman, 2012). Naturally, text mining can be a means to reduce the biocuration effort and enhance the content of biological databases in a number of ways, from selecting potential facts to prioritizing sentences, paragraphs and documents or by annotating a text before it is curated.

Integration

Text mining evidence can be combined with other experimental and computational results both in low-throughput and high-throughput fashion. For example, gene expression data can be gathered from published microarray data sets, mentions of expression in scientific literature and internal experimental sources (Caviness *et al.*, 2011). In such cases, text-mined results need to be organized with the same identifiers as the other data sources, such as HUGO gene symbols for gene-centric

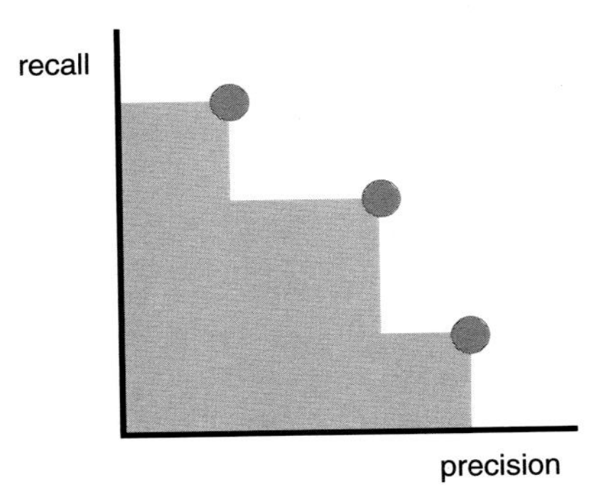

Figure A.6 The performance of a well-behaved suite of strategies for a particular text mining task should follow a downward trend in the precision-recall space. Low recall/high precision will be preferred when less curation time is available. When more curation time is available, high recall/low precision will be preferred. In between these two strategies, other middle-ground strategies are possible.

analyses. These identifiers can also be semantic web URIs that are linked to external and internal resources.

Integration often requires that text-mined results be encoded into an appropriate score or annotation. For example, a simple binary code can be used with TRUE for facts that have been mentioned in a corpus and FALSE for facts that have not. Alternatively, the number of times a fact has been mentioned in a corpus can be used as a score. The more times a fact is mentioned, the more established it is considered.

Visualization and presentation

Text mining results that have been encoded and integrated with results from other experimental and computational sources become another data dimension that can be presented with common visualization tools used in drug discovery. The presentation of raw text mining results, on the other hand, has its own visualization challenges, especially when trying to facilitate the work of curators. As has been mentioned, curation environments can make the task of sifting through text mining results easier.

Text mining can also be used to enhance the visualization of biomedical documents. Documents can be pre-processed with text mining to identify facts, which then are highlighted and linked to other sources of information, such as reference databases and encyclopedias. This markup enriches the reading experience by allowing the reader quick identification of important facts, as can be seen in the PDF reader Utopia Documents (Attwood *et al.*, 2010) or the Reflect internet browser plug-in (Pafilis *et al.*, 2009). An example developed in the pharmaceutical industry is the Ultralink from Novartis (Romacker *et al.*, 2006).

Applications of text mining

Dimensions

Pharmaceutical applications of text mining can be described along a number of dimensions.

Easy and hard questions

Text mining challenges have repeatedly shown that some text mining tasks are harder than others. For example, algorithms that map abbreviations or recognize mutations have reported more success than algorithms that identify events or synonymy.

Obviously, the more structured a source of content, the easier it is to mine. It is easier to find data about the outcome of a trial if it is located in the results section of a structured abstract. Similarly, it is easier to find information about providers of reagents if the search can be narrowed to the methods section of an article. Besides the structure of the document, other factors can make a difference, such as the variability of the language and vocabulary. For example, the yeast gene nomenclature is simpler than that for human genes (Seringhaus *et al.*, 2008). "[...] yeast has a smaller genome than mouse or human, with fewer gene names, and the names are shorter than those of other species: one word on average. In addition, the yeast lexicon listed fewer synonyms on average" (Morgan *et al.*, 2008). This is reflected in the performance of algorithms that extract and normalize yeast and human gene names. Even gene recognition is harder in full text than in abstracts (Lu *et al.*, 2011).

Another factor that affects the difficulty of a text mining task is the frequency of relevant facts that appear in the text. Any text mining algorithm produces a certain rate of false positives, often regardless of whether there are any facts to be extracted in the text. Thus, searching for facts in a set of documents in which they are very unlikely to exist ends up just generating many false positives. This is analogous to looking for very few needles in a very large haystack (Rodriguez-Esteban, 2009; Rodriguez-Esteban and Loging, 2013).

The requirements of drug discovery questions are also important determinants of the difficulty of a text mining task. In particular, the minimum recall and precision requested can be strong constraints. For example, when constructing an interaction network around an important target, recall is very important. However, identifying each of the interactions more than once may not be necessary. Some tasks, on the other hand, require only the generation of a number of suggestions or "good ideas" to be followed up later by humans. For example, when creating a list of promising drug concepts (or targets) in a disease area. Because this task is less technically constrained, it shifts the attention towards understanding the disease and the properties that make a drug concept attractive.

In systems biology and pharmacokinetic modeling, model parameters are often created based on literature mentions of experimental measurements and kinetic values. For building such models the priority is gathering values that are ballpark figures within reasonable effort and time rather than exhausting every source. The Pareto principle (or 80–20 rule) applies in such cases so that a significant number of results can be gathered by covering the most common patterns. The Pareto principle also applies, for example, when building an *ad hoc* vocabulary. It is easy to collect the most common terms, but hard to cover the long tail of rare terms.

Open-ended and closed-ended questions

One way to classify a text mining question is to ask whether it is an "open-ended" or a "closed-ended" question. By closed-ended question I mean a question that has a precise definition and whose answer is straightforward to interpret. By open-ended question I mean a question that might be answered through an undefined number of indirect types of evidence. Of course, there is a continuum between open-ended and closed-ended questions. A clearly closed-ended question would be "what is the molecular weight of nitrogen?" A more open-ended question could involve finding all drug–drug interactions mentioned in the literature for a given drug. A yet more open-ended question would be to identify reliable clinical biomarkers for a drug or to propose promising targets to start a drug development project. The latter question may require looking at different aspects of biology that make a target a good candidate. Because there is no universal agreement on what a "good" target is, by accumulating different types of evidence and integrating them with other biological data sets it is possible to build a case around certain target candidates. A short list of such candidates can then be further analyzed in a "fishing expedition" or target hunting exercise in which a list of targets is split across the members of a group for in-depth analysis of each individual target. Thus, open-ended questions leave more room for interpretation and produces results that are less univocal.

Discovery questions

For some, text mining has a major weakness as a drug discovery tool in comparison to other computational biology approaches: it cannot produce novel findings. While this is literally true, there are two important objections to this argument. First, due to high publication rates, disease scientists are never aware of everything that has been published in their own fields. Thus, findings produced by text mining can appear "novel" to anyone who is not aware of them. The second objection is that combining information from different sources can lead to what is called literature-based discoveries.

The main method used for literature-based discovery is transitive inference (Weeber *et al.*, 2005; Weeber, 2007). Transitive inference borrows its strategy from deductive logic. The simplest form of transitive inference is the ABC model (Swanson, 1986), which states that if A is related to B and B related to C, then

A might be related to C, even if there is no published proof of that (yet). Through statistical analysis based on the ABC model, it is possible to anticipate with a certain degree of confidence new protein relationships years before they appear in the literature (van Haagen *et al.*, 2009; Frijters *et al.*, 2010).

A useful extension of the ABC model for pharmaceutical research is the closed discovery paradigm (Weeber *et al.*, 2001). In closed discovery, two elements, one upstream (A) and another downstream (C), can be linked through intermediate elements (Bs). The Bs are the overlap between the set of all elements downstream from A and the set of all elements upstream from C. The process of finding the Bs can be simplified by doing the steps sequentially. First, identifying all downstream effects of A and, second, checking whether any one of those downstream effects are upstream from C.

The ABC model can also be generalized into "discovery browsing," in which connections are iteratively explored with the goal of connecting A and C through a sequence of intermediate Bs (Wilkowski *et al.*, 2011).

Practical examples

A number of applications of text mining in pharmaceutical companies have been described in the scientific literature. One typical application is the construction of gene- and drug-centric databases populated with relationships extracted from text (Gomes *et al.*, 2006; Agarwal and Searls, 2008; Kumar, 2011; Tari *et al.*, 2011). The types of relationships covered by these databases vary depending on the focus of the database. Some of these databases also further integrate text mining data with bioinformatics, cheminformatics and competitive intelligence data sources. These integrated databases can help review and visualize the space of potential drug targets and competitive landscape (Loging *et al.*, 2007; Harland *et al.*, 2009; Campbell *et al.*, 2012) and identify research trends (Rajpal *et al.*, 2011).

Other pharmaceutical applications of text mining have covered areas such as drug safety (Gomes *et al.*, 2006; Davis *et al.*, 2013), systems biology (Cho *et al.*, 2006; Gomes *et al.*, 2006), interaction networks (Murray *et al.*, 2010) and analysis of scientific trends (Agarwal and Searls, 2009). Two other examples of text mining applications are presented here.

Patent mining of animal models

As mentioned, the USPTO has made available the complete set of US accepted patents and patent applications through Google Patents. The text of the patents is provided alongside additional files detailing approval history, figures, sequences, mathematical formulations and other types of information. The number of figure images contained in this resource is very large. The combined file size of the images exceeds 1 TB in zipped format. One way to sort such large number of figures is by classifying them according to their properties using machine learning methods. Figure image features can be created based on the figure's color histograms and shape and texture descriptors. Features can also be created from the figure captions and legends and by examining the text around the location where the figures are cited.

A particular request from discovery scientists involved the identification of patent figures that depicted response curves to treatment in a particular animal model of importance for a disease. To address this problem, all USPTO patents mentioning the disease were gathered and their XML parsed. Figure legends in the patents' text were identified using a set of regular expressions (rules). Then, a training and test set was created from a random sample of the figures. A simple HTML interface was created to evaluate the random sample. Users were presented with the randomly selected figures and asked for their relevance in a binary yes/no fashion. The resulting set was used as a gold standard.

From previous work (Rodriguez-Esteban and Iossifov, 2009) the author had learned that figure legends may be the most informative features for figure classification. Thus, features describing figure legends were created using the bag-of-words method. A support vector machine (SVM) algorithm was trained and tested on features derived from the gold standard. Once the algorithm was trained and tested it was applied to the set of all figures. The algorithm produced a score for each figure according to its relevance and these scores were used to rank the figures. The figures with the highest score were further examined for their relevance.

Literature holes

Using text mining tools it is possible to map the entire literature about a specific subject. This allows the identification of "holes" in the literature, meaning areas that have been neglected but that could be nonetheless of interest. In other words, text mining can produce a map of a portion of biomedicine and from that derive the "terra incognita" that has not been explored yet.

A practical example of such an approach is the analysis of all the knockout phenotypes published for a reference animal model. Such analysis can produce an overview of the therapeutic ideas that have been considered in diseases that have such an animal model as a reference. Mentions of knockout experiments can be captured using rule-based templates to identify relationships at the sentence level between gene knockout experiments and changes in the animal model's phenotype. If the animal model and the phenotype are well-defined, the mining can be done comprehensively, covering conference abstracts, full-text articles, and patents.

Identified knockout genes can then be studied further by overlaying them on interaction networks and pathway diagrams and seeking other genes in the same network and pathway neighborhoods that have not been studied; for example, by using the ABC model explained in the Discovery section of this chapter. If proteins A and C in a signaling cascade have been linked to a disease and protein B has not, then it could be fruitful to investigate B. Thus, text mining can pinpoint the scientific "holes" that exist in established biological areas that have been thoroughly investigated.

Text mining within the pharmaceutical organization

Text mining within pharmaceutical companies often sits at the crossroads of computational biology and information science. From the point of view of information scientists, text mining appears like an enhanced expert search that takes advantage of innovations from information extraction and NLP (Roberts and Hayes, 2008). From the point of view of computational biology, text mining is a high-throughput *in silico* analysis of biomedical data on par with other computational biology fields such as genomics and systems biology (see also the discussion on pp. 365–366 of Hersh, 2009). Indeed, some text mining exercises are closer to expert searches because they focus on finding concrete and well-specified pieces of knowledge ("closed-ended" questions). Other text mining applications, however, involve large sets of results which are integrated alongside other high-throughput omics data sets (Loging *et al.*, 2007; Harland *et al.*, 2009; Kumar, 2011; Campbell *et al.*, 2012). In these cases, the information mined from text can be comparable to that produced by other high-throughput methods such as microarray experiments, DNA sequencing or protein–protein interaction screenings. Deciding whether any single text mining exercise belongs to the discipline of information science or of computational biology might be a rather unnecessary and futile exercise.

Text mining tools have shown their usefulness in other parts of the pharmaceutical organization beyond R&D, such as medical affairs, marketing and public relations; with applications such as identifying medical experts and key opinion leaders, post-marketing safety monitoring and analyzing social media. However, a corporation needs to organize its resources according to a structure in which individuals belong to groups with specific sets of goals. Thus, when text mining skills are located in the research division, as often happens, text mining applications for the development, marketing and medical divisions are far less emphasized. Ideally, text mining should be a global function within a pharmaceutical company.

Unlike other areas of bioinformatics, text mining has been adopted mainly by large- and medium-sized pharmaceutical companies. Such companies have larger needs for text mining and have the financial means to support the tools and licenses to mine the scientific literature. New business models, already being discussed, could reduce these financial requirements, for example by allowing a "text mining at the source" or "pre-mining" which pushes parts of the text mining process to the content providers' side. The content providers thus would deliver sets of results already text-mined to some extent and according to the consumers' needs. This would eliminate the need for the consumer to create large corpora, yielding much smaller literature needs and lower costs acceptable to smaller companies.

References

Agarwal, P. and Searls, D. B. Literature mining in support of drug discovery. *Briefings in Bioinformatics*. 2008;9(6):479–492.

Agarwal, P. and Searls, D. B. Can literature analysis identify innovation drivers in drug discovery? *Nature Reviews Drug Discovery*. 2009;8(11):865–878.

Aronson, A. R. Effective mapping of biomedical text to the UMLS Metathesaurus: The MetaMap program. *Proceedings of the AMIA Symposium*. 2001;17–21.

Attwood, T. K., Kell, D. B., McDermott, P., *et al.* Utopia documents: Linking scholarly literature with research data. *Bioinformatics*. 2010;26(18):i568–574.

Bui, D. D. and Zeng-Treitler, Q. Learning regular expressions for clinical text classification. *Journal of the American Medical Information Association*. 2014;21(5):850–857.

Burger, J. D., Doughty, E., Khare, R., *et al.* Hybrid curation of gene-mutation relations combining automated extraction and crowdsourcing. *Database (Oxford)*. 2014;2014(pii):bau094.

Campbell, S. J., Gaulton, A., Marshall, J., *et al.* Visualizing the drug target landscape. *Drug Discovery Today*. 2012;17(Suppl):S3–15.

Caviness, G., Jiang, X., Qi, Z. and Rodriguez-Esteban, R. An Integrative Approach on Genomic, Genetic and Text Mining Based Information for Psoriasis Target Identification. Non-Clinical Biostatistics Conference, Cambridge, MA, 2011.

Chang, J. T., Schütze, H. and Altman, R. B. Creating an online dictionary of abbreviations from MEDLINE. *Journal of the American Medical Information Association*. 2002;9(6):612–620.

Chen, L. and Friedman, C. Extracting phenotypic information from the literature via natural language processing. *Studies in Health Technology Information*. 2004;107(Pt 2):758–762.

Chen, L., Liu, H. and Friedman, C. Gene name ambiguity of eukaryotic nomenclatures. *Bioinformatics*. 2005;21(2):248–256.

Chinchor, N. MUC-4 evaluation metrics. Proceedings of the 4th Conference on Message Understanding, 1992.

Chinchor, N. Overview of MUC-7. 7th Message Understanding Conference (MUC-7), 1998.

Cho, C. R., Labow, M., Reinhardt, M., van Oostrum, J. and Peitsch, M. C. The application of systems biology to drug discovery. *Current Opinion in Chemical Biology*. 2006;10(4):294–302.

Cohen, A. M., Hersh, W. R., Dubay, C. and Spackman, K. Using co-occurrence network structure to extract synonymous gene and protein names from MEDLINE abstracts. *BMC Bioinformatics*. 2005;6:103.

Cohen, K. B., Dolbey, A. E., Acquaah-Mensah, G. K. and Hunter, L. Contrast and variability in gene names. *ACL Workshop on Natural Language Processing in the Biomedical Domain*. 2002;14–20.

Cokol, M. and Rodriguez-Esteban, R. Visualizing evolution and impact of biomedical fields. *Journal of Biomedical Information*. 2008;41(6):1050–1052.

Comeau, D.C., Islamaj Doğan, R., Ciccarese, P., *et al.* BioC: a minimalist approach to interoperability for biomedical text processing. *Database (Oxford)*. 2013;2013:bat064. doi: 10.1093/database/bat064

Czarnecki, J., Nobeli, I., Smith, A. M. and Shepherd, A. J. A text-mining system for extracting metabolic reactions from full-text articles. *BMC Bioinformatics*. 2012;13:172.

Davis, A. P., Wiegers, T. C., Roberts, P. M., *et al.* A CTD–Pfizer collaboration: manual curation of 88,000 scientific articles text mined for drug–disease and drug–phenotype interactions. *Database (Oxford)*. 2013;2013:bat080.

Ding, J., Berleant, D., Nettleton, D. and Wurtele, E. Mining MEDLINE: Abstracts, sentences, or phrases? *Pacific Symposia on Biocomputing*. 2002:326–337.

Doğan, R. I., Leaman, R. and Lu, Z. NCBI disease corpus: A resource for disease name recognition and concept normalization. *Journal of Biomedical Information*. 2014;47:1–10.

Franceschini, A., Szklarczyk, D., Frankild, S., *et al.* STRING v9.1: Protein–protein interaction networks, with increased coverage and integration. *Nucleic Acids Research*. 2013;41(Database issue):D808–815.

Friedman, C., Kra, P., Yu, H., Krauthammer, M. and Rzhetsky, A. GENIES: A natural-language processing system for the extraction of molecular pathways from journal articles. *Bioinformatics*. 2001;17(Suppl 1):S74–82.

Frijters, R., van Vugt, M., Smeets, R., *et al.* Literature mining for the discovery of hidden connections between drugs, genes and diseases. *PLoS Computers and Biology*. 2010;6(9).

Fukuda, K., Tamura, A., Tsunoda, T. and Takagi, T. Toward information extraction: Identifying protein names from biological papers. *Pacific Symposia on Biocomputing*. 1998;707–718.

Funk, C., Baumgartner, W. Jr, Garcia, B., *et al.* Large-scale biomedical concept recognition: An evaluation of current automatic annotators and their parameters. *BMC Bioinformatics*. 2014;15(1):59.

Furrer, L., Clematide, S., Marques, H., *et al.* Collection-wide extraction of protein–protein interactions. 6th International Symposium on Semantic Mining in Biomedicine (SMBM). Aveiro, Portugal, October 2014.

Gerner, M., Nenadic, G. and Bergman, C. M. LINNAEUS: A species name identification system for biomedical literature. *BMC Bioinformatics*. 2010;11:85.

Gomes, B., Hayes, W. and Podowski, R. M. Text mining. In *In Silico Technologies in Drug Target Identification and Validation* (pp. 153–194), ed. D. Leon and S . Markel. Boca Raton, FL: CRC Press, 2006.

Grishman, R. and Sundheim, B. Message Understanding Conference-6: A brief history. *16th Conference on Computational Linguistics*. 1996(1):466–471.

van Haagen, H. H., 't Hoen, P. A., Botelho Bovo, A., *et al.* Novel protein–protein interactions inferred from literature context. *PLoSONE*. 2009;4(11):e7894.

Hakenberg, J., Schmeier, S., Kowald, A., Klipp, E. and Leser, U. Finding kinetic parameters using text mining. *OMICS*. 2004;8(2):131–152.

Hakenberg, J., Gerner, M., Haeussler, M., *et al.* The GNAT library for local and remote gene mention normalization. *Bioinformatics*. 2011;27(19):2769–2771.

Harland, L. and Gaulton, A. Drug target central. *Expert Opinion in Drug Discovery*. 2009;4(8):857–872.

Hatzivassiloglou, V., Duboué, P. A. and Rzhetsky, A. Disambiguating proteins, genes, and RNA in text: A machine learning approach. *Bioinformatics*. 2001;17(Suppl 1):S97–106.

Hatzivassiloglou, V. and Weng, W. Learning anchor verbs for biological interaction patterns from published text articles. *International Journal of Medical Information*. 2002;67(1–3):19–32.

Herrero-Zazo, M., Segura-Bedmar, I., Martínez, P. and Declerck, T. The DDI corpus: An annotated corpus with pharmacological substances and drug–drug interactions. *Journal of Biomedical Information*. 2013;46(5):914–920.

Hersh, W. R. *Information Retrieval: A Health and Biomedical Perspective* (3rd edition). New York, NY: Springer, 2009.

Hersh, W. R., Bhupatiraju, R. T., Ross, L., *et al.* Enhancing access to the Bibliome: The TREC 2004 Genomics Track. *Journal of Biomedical Discovery and Collaboration*. 2006;1:3.

Hirschman, L., Burns, G. A., Krallinger, M., *et al.* Text mining for the biocuration workflow. *Database (Oxford)*. 2012;2012:bas020.

Hirschman, L., Morgan, A. A. and Yeh, A. S. Rutabaga by any other name: Extracting biological names. *Journal of Biomedical Information*. 2002;35(4):247–259.

Hirst, G. Near-synonymy and the structure of lexical knowledge. *AAAI Symposium on Representation and Acquisition of Lexical Knowledge: Polysemy, Ambiguity, and Generativity*. 1995:51–56.

Hu, Y., Hines, L. M., Weng, H., *et al.* Analysis of genomic and proteomic data using advanced literature mining. *Journal of Proteome Research*. 2003;2(4):405–412.

Hu, Z. Z., Narayanaswamy, M., Ravikumar, K. E., Vijay-Shanker, K., Wu, C. H. Literature mining and database annotation of protein phosphorylation using a rule-based system. *Bioinformatics*. 2005;21(11):2759–2765.

Hunter, L., Lu, Z., Firby, J., *et al.* OpenDMAP: An open source, ontology-driven concept analysis engine, with applications to capturing knowledge regarding protein transport, protein interactions and cell-type-specific gene expression. *BMC Bioinformatics*. 2008;9:78.

Iossifov, I., Rodriguez-Esteban, R., Mayzus, I., Millen, K. J. and Rzhetsky, A. Looking at cerebellar malformations through text-mined interactomes of mice and humans. *PLoS Computational Biology*. 2009;5(11):e1000559.

Jamieson, D. G., Roberts, P. M., Robertson, D. L., Sidders, B. and Nenadic, G. Cataloging the biomedical world of pain through semi-automated curation of molecular interactions. *Database (Oxford)*. 2013;2013:bat033.

Jessop, D. M., Adams, S. E., Willighagen, E. L., Hawizy, L. and Murray-Rust, P. OSCAR4: A flexible architecture for chemical text-mining. *Journal of Cheminformation*. 2011;3(1):41.

Jimeno, A., Jimenez-Ruiz, E., Lee, V., *et al.* Assessment of disease named entity recognition on a corpus of annotated sentences. *BMC Bioinformatics*. 2008;9(Suppl 3):S3.

Jonnalagadda, S. R. and Topham, P. NEMO: Extraction and normalization of organization names from PubMed affiliations. *Journal of Biomedical Discovery Collaboration*. 2010;5:50–75.

Jonquet, C., Shah, N. H. and Musen, M. A. The open biomedical annotator. *Summit on Translational Bioinformation*. 2009;2009:56–60.

Kabiljo, R., Clegg, A. B. and Shepherd, A. J. A realistic assessment of methods for extracting gene/protein interactions from free text. *BMC Bioinformatics*. 2009;10:233.

Kageura, K. and Umino, B. Methods of automatic term recognition – A review. *Terminology*. 1996;3:259–289.

Kim, J. D., Ohta, T. and Tsujii, J. Corpus annotation for mining biomedical events from literature. *BMC Bioinformatics*. 2008;9:10.

Köhler, S., Doelken, S. C., Mungall, C. J., *et al.* The Human Phenotype Ontology project: Linking molecular biology and disease through phenotype data. *Nucleic Acids Res*. 2014;42(Database issue):D966–974.

Krallinger, M., Leitner, F., Rabal, O., *et al.* Overview of the chemical compound and drug name recognition (CHEMDNER) task. *Proceedings of the Fourth BioCreative Challenge Evaluation Workshop*. 2013;2:2–33.

Krallinger, M., Leitner, F., Rodriguez-Penagos, C. and Valencia, A. Overview of the protein–protein interaction annotation extraction task of BioCreative II. *Genome Biology*. 2008;9(Suppl 2):S4.

Krauthammer, M. and Nenadic, G. Term identification in the biomedical literature *Journal of Biomedical Information*. 2004;37:512–526.

Kumar, V. Omics and literature mining. *Methods in Molecular Biology*. 2011;719:457–477.

Leaman, R. and Gonzalez, G. BANNER: An executable survey of advances in biomedical named entity recognition. *Pacific Symposia on Biocomputing*. 2008:652–663.

Leaman, R., Islamaj Dogan, R. and Lu, Z. DNorm: Disease name normalization with pairwise learning to rank. *Bioinformatics*. 2013;29(22):2909–2917.

Li, C. X., Wang, R. J., Chen, P., Huang, H. and Su, Y. R. Interaction relation ontology learning. *Journal of Computational Biology*. 2014;21(1):80–88.

Li, Y. and Agarwal, P. A pathway-based view of human diseases and disease relationships. *PLoS ONE*. 2009;4(2):e4346. doi: 10.1371/journal.pone.0004346.

Liang, G. and Hai, T. Characterization of human activating transcription factor 4, a transcriptional activator that interacts with multiple domains of cAMP-responsive element-binding protein (CREB)-binding protein. *Journal of Biological Chemistry*. 1997;272(38):24088–24095.

Liu, H., Hu, Z. Z., Torii, M., Wu, C. and Friedman, C. Quantitative assessment of dictionary-based protein named entity tagging. *Journal of the American Medical Information Association*. 2006;13(5):497–507.

Liu, W., Islamaj Dogan, R., Kim, S., *et al.* Author name disambiguation for PubMed. *Journal of the Association for Information Science and Technology*. 2014;65(4):765–781.

Loging, W., Harland, L. and Williams-Jones, B. High-throughput electronic biology: Mining information for drug discovery. *Nature Reviews Drug Discovery*. 2007;6(3):220–230.

Lovis, C., Michel, P. A., Baud, R. and Scherrer, J. R. Word segmentation processing: A way to exponentially extend medical dictionaries. *Medinfo*. 1995;8(Pt 1):28–32.

Lu, Z. and Hirschman, L. Biocuration workflows and text mining: Overview of the BioCreative 2012 Workshop Track II. *Database (Oxford)*. 2012;2012:bas043.

Lu, Z., Kao, H. Y., Wei, C. H., *et al.* The gene normalization task in BioCreative III. *BMC Bioinformatics*. 2011;12(Suppl 8):S2.

Malhotra, A., Younesi, E., Gurulingappa, H. and Hofmann-Apitius, M. "HypothesisFinder": A strategy for the detection of speculative statements in scientific text. *PLoS Computational Biology*. 2013;9(7):e1003117.

Mani, I., Hu, Z., Jang, S. B., *et al.* Protein name tagging guidelines: Lessons learned. *Comparative and Functional Genomics*. 2005;6(1–2):72–76.

Midford, P. E., Dececchi, T. A., Balhoff, J. P., *et al.* The vertebrate taxonomy ontology: A framework for reasoning across model organism and species phenotypes. *Journal of Biomedical Semantics*. 2013;4(1):34.

Mihăilă, C., Batista-Navarro, R. T. and Ananiadou, S. Analysing entity type variation across biomedical subdomains. *Proceedings of the Third Workshop on Building and Evaluating Resources for Biomedical Text Mining* (BioTxtM 2012). 2012:1–7.

Mihăilă, C., Ohta, T., Pyysalo, S. and Ananiadou, S. BioCause: Annotating and analysing causality in the biomedical domain. *BMC Bioinformatics*. 2013;14:2.

Milward, D., Bjäreland, M., Hayes, W., *et al.* Ontology-based interactive information extraction from scientific abstracts. *Comparative and Functional Genomics*. 2005;6(1–2):67–71.

Milward, D., Blaschke, C., Neefs, J., *et al.* Flexible text mining strategies for drug discovery. *Proceedings of the Second International Symposium on Semantic Mining in BioMedicine (SMBM)*. 2006:101–104.

Miwa, M., Ohta, T., Rak, R., *et al.* A method for integrating and ranking the evidence for biochemical pathways by mining reactions from text. *Bioinformatics*. 2013a;29(13):i44–52.

Miwa, M., Pyysalo, S., Ohta, T. and Ananiadou, S. Wide coverage biomedical event extraction using multiple partially overlapping corpora. *BMC Bioinformatics*. 2013b;14:175.

Morgan, A. A., Lu, Z., Wang, X., *et al.* Overview of BioCreative II gene normalization. *Genome Biology*. 2008;9(Suppl 2):S3.

Murray, B. S., Choe, S. E., Woods, M., Ryan, T. E. and Liu, W. An *in silico* analysis of microRNAs: mining the miRNAome. *Molecular bioSystems*. 2010;6(10):1853–1862.

Neves, M. L., Carazo, J. M. and Pascual-Montano, A. Moara: A Java library for extracting and normalizing gene and protein mentions. *BMC Bioinformatics*. 2010;11:157.

Pafilis, E., Frankild, S. P., Fanini, L., *et al.* The SPECIES and ORGANISMS resources for fast and accurate identification of taxonomic names in text. *PLoS ONE*. 2013;8(6):e65390.

Pafilis, E., O'Donoghue, S. I., Jensen, L. J., *et al.* Reflect: Augmented browsing for the life scientist. *Nature Biotechnology*. 2009;27(6):508–510.

Pustejovsky, J., Castaño, J., Cochran, B., Kotecki, M. and Morrell, M. Automatic extraction of acronym-meaning pairs from MEDLINE databases. *Studies in Health Technology Information*. 2001;84(Pt 1):371–375.

Pyysalo, S., Airola, A., Heimonen, J., *et al.* Comparative analysis of five protein–protein interaction corpora. *BMC Bioinformatics*. 2008;9(Suppl 3):S6.

Pyysalo, S., Ohta, T., Rak, R., *et al.* Overview of the ID, EPI and REL tasks of BioNLP Shared Task 2011. *BMC Bioinformatics*. 2012;13(Suppl 11):S2.

Rajpal, D. K., Kumar, V. and Agarwal, P. Scientific literature mining for drug discovery: A case study on obesity. *Drug Development Research*. 2011;72:201–208.

Rebholz-Schuhmann, D., Marcel, S., Albert, S., *et al.* Automatic extraction of mutations from Medline and cross-validation with OMIM. *Nucleic Acids Research*. 2004;32(1):135–142.

Rebholz-Schuhmann, D., Kafkas, S., Kim, J. H., *et al.* Evaluating gold standard corpora against gene/protein tagging solutions and lexical resources. *Journal of Biomedical Semantics*. 2013;4(1):28. doi: 10.1186/20411480428.

Rinaldi, F., Clematide, S., Marques, H., *et al.* OntoGene web services for biomedical text mining. *BMC Bioinformatics*. 2014;15(Suppl 14):S6.

Rinaldi, F., Kappeler, T., Kaljurand, K., *et al.* OntoGene in BioCreative II. *Genome Biology*. 2008;9(Suppl 2):S13.

Rocktäschel, T., Weidlich, M. and Leser, U. ChemSpot: A hybrid system for chemical named entity recognition. *Bioinformatics*. 2012;28(12):1633–1640.

Roberts, P. M. and Hayes, W. S. Information needs and the role of text mining in drug development. *Pacific Symposia on Biocomputing*. 2008:592–603.

Rodriguez-Esteban, R. Biomedical text mining and its applications. *PLoS Computational Biology*. 2009; 5(12):e1000597.

Rodriguez-Esteban, R. and Iossifov, I. Figure mining for biomedical research. *Bioinformatics*. 2009;25(16):2082–2084.

Rodriguez-Esteban, R., Iossifov, I. and Rzhetsky, A. Imitating manual curation of text-mined facts in biomedicine. *PLoS Computational Biology*. 2006;2(9):e118.

Rodriguez-Esteban, R. and Loging, W. T. Quantifying the complexity of medical research. *Bioinformatics*. 2013;29(22):2918–2924.

Rodriguez-Esteban, R., Roberts, P. M. and Crawford, M. E. Identifying and classifying biomedical perturbations in text. *Nucleic Acids Research*. 2009;37(3):771–777.

Romacker, M., Grandjean, N., Parisot, P., *et al.* The UltraLink: An expert system for contextual hyperlinking in knowledge management. In *Computer Applications in Pharmaceutical Research and Development*, ed. S. Ekins. Hoboken, NJ: Wiley, 2006.

Rzhetsky, A. and Evans, J. A. War of ontology worlds: Mathematics, computer code, or Esperanto? *PLoS Computational Biology*. 2011;7(9):e1002191.

Rzhetsky, A., Koike, T., Kalachikov, S., *et al.* A knowledge model for analysis and simulation of regulatory networks. *Bioinformatics*. 2000;16(12):1120–1128.

Saric, J., Jensen, L. J., Ouzounova, R., Rojas, I. and Bork, P. Extraction of regulatory gene/protein networks from Medline. *Bioinformatics*. 2006 ;22(6):645–650.

Sekimizu, T., Park, H. S. and Tsujii, J. Identifying the interaction between genes and gene products based on frequently seen verbs in Medline abstracts. *Genome Information Series Workshop Genome Information*. 1998;9:62–71.

Seringhaus, M. R., Cayting, P. D. and Gerstein, M. B. Uncovering trends in gene naming. *Genome Biology*. 2008;9(1):401.

Shatkay, H., Pan, F., Rzhetsky, A. and Wilbur, W. J. Multi-dimensional classification of biomedical text: Toward automated, practical provision of high-utility text to diverse users. *Bioinformatics*. 2008;24(18):2086–2093.

Smith, C. L., Goldsmith, C. A. and Eppig, J. T. The Mammalian Phenotype Ontology as a tool for annotating, analyzing and comparing phenotypic information. *Genome Biology*. 2005;6(1):R7.

Su, L. K., Burrell, M., Hill, D. E., *et al.* APC binds to the novel protein EB1. *Cancer Research*. 1995;55(14):2972–2977.

Swanson, D. R. Fish oil, Raynaud's syndrome, and undiscovered public knowledge. *Perspectives in Biology and Medicine*. 1986;30(1):7–18.

Takeda, S., Onishi, Y., Kawamura, T. and Maeda, H. Clinical spectrum of pulmonary inflammatory myofibroblastic tumor. *Interactive Cardiovascular and Thoracic Surgery*. 2008;7(4):629–633.

Tanenblatt, M., Coden, A. and Sominsky, I. The ConceptMapper approach to named entity recognition. Proceedings of Seventh International Conference on Language Resources and Evaluation. 2010.

Tari, L., Küntzer, J., Patel, J., *et al.* Mining gene-centric relationships from literature to support drug discovery. IEEE International Conference on Bioinformatics and Biomedicine. 2011.

Thompson, P., Nawaz, R., McNaught, J. and Ananiadou, S. Enriching a biomedical event corpus with meta-knowledge annotation. *BMC Bioinformatics*. 2011;12:393.

Tikk, D., Solt, I., Thomas, P. and Leser, U. A detailed error analysis of 13 kernel methods for protein–protein interaction extraction. *BMC Bioinformatics*. 2013;14:12.

Tomanek, K., Wermter, J. and Hahn, U. Sentence and token splitting based on conditional random fields. Proceedings of the 10th Conference of the Pacific Association for Computational Linguistics (PACLING), Melbourne, Australia, 2007.

Torii, M., Li, G., Li, Z., *et al.* RLIMS-P: An online text-mining tool for literature-based extraction of protein phosphorylation information. *Database (Oxford)*. 2014;2014.

Torvik, V. I. and Smalheiser, N. R. Author name disambiguation in MEDLINE. *ACM Transactions in Knowledge Discovery Data*. 2009;3(3).pii: 11.

Van Landeghem, S., Hakala, K., Rönnqvist, S., *et al.* Exploring biomolecular literature with EVEX: Connecting genes through events, homology, and indirect associations. *Advances in Bioinformatics*. 2012;2012:582765.

Varadharajalu, A., Liu, W. and Wong, W. Author name disambiguation for ranking and clustering pubmed data using NetClus. In: *AI 2011: Advances in Artificial Intelligence*. Lecture Notes in Computer Science. 2011(7106):152–161.

Weeber, M. Drug discovery as an example of literature-based discovery. In *Computational Discovery of Scientific Knowledge* (pp. 290–306), ed. S. Džeroski and L. Todorovski. New York, NY: Springer, 2007.

Weeber, M., Klein, H., Berg, L. and Vos, R. Using concepts in literature-based discovery: Simulating Swanson's Raynaud–fish oil and migraine–magnesium discoveries. *Journal of the American Society for Information Sciences*. 2001;52(7):548–557.

Weeber, M., Kors, J. A. and Mons, B. Online tools to support literature-based discovery in the life sciences. *Briefings in Bioinformation*. 2005;6(3):277–286.

Wei, C. H., Harris, B. R., Kao, H. Y. and Lu, Z. tmVar: A text mining approach for extracting sequence variants in biomedical literature. *Bioinformatics*. 2013;29(11):1433–1439.

Wilkowski, B., Fiszman, M., Miller, C. M., *et al.* Graph-based methods for discovery browsing with semantic predications. *AMIA Annual Symposium Proceedings.* 2011;2011:1514–1523.

Wren, J. D. and Garner, H. R. Heuristics for identification of acronym-definition patterns within text: Towards an automated construction of comprehensive acronym-definition dictionaries. *Methods of Information in Medicine*. 2002;41(5):426–434.

Wu, C., Schwartz, J. and Nenadic, G. PathNER: A tool for systematic identification of biological pathway mentions in the literature. *BMC Systems Biology*. 2013;7(Suppl 3):S2.

Yang, Z., Li, Y., Cai, J. and Nyberg, E. QUADS: Question answering for decision support. *Proceedings of the 37th international ACM SIGIR conference on Research & development in information retrieval (SIGIR '14)*. 2014:375–384.

Xu, R. and Wang, Q. Large-scale extraction of accurate drug–disease treatment pairs from biomedical literature for drug repurposing. *BMC Bioinformatics*. 2013;14:181.

Xu, Y., Hua, J., Ni, Z., *et al.* Anatomical entity recognition with a hierarchical framework augmented by external resources. *PLoS ONE*. 2014;9(10):e108396.

Zhang, L., Berleant, D., Ding, J. and Wurtele, E. S. Automatic extraction of biomolecular interactions: an empirical approach. *BMC Bioinformatics*. 2013;14:234.

Appendix II Open source tools and public data sources

Yirong Wang and William T. Loging

As technology evolves, the drug discovery science is quickly becoming data science rather than purely biological sciences or chemistry in the traditional sense. Scientists are generating experimental and clinical data at ever-faster paces, many of which become publicly available. How to best use these data and realize the full potential of the data becomes a continuous topic throughout the drug discovery process.

In this section, rather than provide an exhaustive list of the resources available to the scientist, we focus on the few widely used (free or commercial) data sources and analysis tools.

Literature and textual resources

1. eUtils

The Entrez Programming Utilities (E-utilities, or eUtils) are a set of nine server-side programs that provide a stable interface into the Entrez query and database system at the National Center for Biotechnology Information (NCBI). The E-utilities use a fixed URL syntax that translates a standard set of input parameters into the values necessary for various NCBI software components to search for and retrieve the requested data. The E-utilities are therefore the structured interface to the Entrez system, which currently includes 38 databases covering a variety of biomedical data, including nucleotide and protein sequences, gene records, three-dimensional molecular structures, and the biomedical literature.

2. EMBASE

With extensive international journal and conference coverage, Embase is a key resource for generating systematic reviews, supporting effective evidence-based medicine and drug and medical device tracking. Embase facilitates the clinical decision-making process and allows you to get to market faster, while still ensuring the required drug safety and pharmacovigilance.

3. MeSH

MeSH is the National Library of Medicine's controlled vocabulary thesaurus. It consists of sets of terms naming descriptors in a hierarchical structure that permits searching at various levels of specificity.

Genomics resources

1. OMIM

OMIM is a comprehensive, authoritative compendium of human genes and genetic phenotypes that is freely available and updated daily. OMIM is authored and edited at the McKusick-Nathans Institute of Genetic Medicine, Johns Hopkins University School of Medicine, under the direction of Dr. Ada Hamosh. Its official home is omim.org.

2. dbVar

A database of genomic structural variation including insertions, deletions, duplications, inversions, deletion-insertions, mobile element insertions, translocations, and complex rearrangements.

3. GEO

GEO is a public functional genomics data repository supporting MIAME-compliant data submissions. Array- and sequence-based data are accepted. Tools are provided to help users query and download experiments and curated gene expression profiles.

4. BioMart

A highly customizable data mining tool that allows web/local programming and mining the Ensembl databases.

Functional analysis

1. DAVID Bioinformatics Resources

The **D**atabase for **A**nnotation, **V**isualization and **I**ntegrated **D**iscovery **(DAVID)** v6.7 is an update to the sixth version of the original web-accessible programs. DAVID now provides a comprehensive set of functional annotation tools for investigators to understand the biological meaning behind a large list of genes.

2. BioSystems

The NCBI BioSystems Database provides integrated access to biological systems and their component genes, proteins, and small molecules, as well as literature describing those biosystems and other related data throughout Entrez.

Workflow and other tools

1. Taverna

Taverna is an open-source and domain-independent workflow management system – a suite of tools used to design and execute scientific workflows and aid *in silico* experimentation.

2. Bioconductor

Bioconductor provides tools for the analysis and comprehension of high-throughput genomic data. Bioconductor uses the R statistical programming language, and is open source and open development.

3. Cytoscape

Cytoscape is an open source software platform for visualizing complex networks and integrating these with any type of attribute data. A lot of apps are available for various kinds of problem domains, including bioinformatics, social network analysis, and semantic web.

4. tranSMART

The tranSMART platform is an open-source, community driven knowledge management platform for translational medicine. It is a project that is being collaboratively developed by more than 100 computer scientists and physician scientists from more than 20 organizations from around the world.

Commercial tools (no affiliation)

1. Trialtrove

Trialtrove is the world's most comprehensive real-time source of pharmaceutical clinical trials intelligence, gathering clinical trial information from over 30,000 clinical trial data sources to provide a continually updated reference of clinical trials research in more than 150 countries.

2. BIOVIA

BIOVIA, a leading provider of scientific innovation lifecycle management software, supports industries and organizations that rely on scientific innovation to differentiate themselves. The industry-leading Pipeline Pilot scientific data processing application provides a broad, flexible scientific foundation optimized to integrate the diversity of science, experimental processes and information requirements across the research, development, QA/QC and manufacturing phases of product development. BIOVIA offers capabilities in scientific data management, modeling and simulation, research informatics, laboratory informatics, enterprise quality management, environmental health and safety, and operations intelligence for customers in science-driven industries. Using BIOVIA technology, scientific innovators can access, organize, analyze and share data in unprecedented ways, ultimately enhancing innovation, improving productivity and compliance, reducing costs and accelerating product development from research to manufacturing.

3. Ayasdi Core

Ayasdi Core is an advanced analytics application that helps uncover critical business intelligence from highly complex and growing data sets. It uses the broadest range of algorithms combined with topological data analysis to accelerate the discovery of insights, hidden or previously overlooked by conventional analytical approaches.

Resource	Brief description of resource	URL
Literature and textual resources		
Medline	Primary repository of biomedical literature abstracts	http://pubmed.gov
EMBASE	Repository of biomedical abstracts, including a significant proportion of journals not covered by Medline	
Medical subject headings (MeSH)	US National Library of Medicine's controlled vocabulary used for indexing articles for Medline	www.nlm.nih.gov/mesh/meshhome.html
Unified medical language (UMLS)	Compendium of different biomedical vocabularies and mappings between them	http://umlsinfo.nlm.nih.gov
US Food and Drug Administration (FDA) RSS feeds	Real-time delivery of important safety and regulatory information from the US FDA agency	www.fda.gov/AboutFDA/ContactFDA/StayInformed/RSSFeeds/default.htm
Biochemical pathways and protein interactions		
Kyoto Encyclopedia of Genes and Genomes (KEGG)	Public database of molecular interactions and biochemical pathways	www.genome.jp/kegg/
BioCarta	Public database of biochemical pathways with a community-focused maintenance policy	www.biocarta.com

Resource	Brief description of resource	URL
Ingenuity Pathway Analysis	Commercial pathway suite based on expert curated content	www.ingenuity.com
MetaCore	Commercial pathway suite based on expert curated content	http://lsresearch.thomsonreuters.com/
Pathway Studio	Commercial pathway suite based on expert curated content	www.elsevier.com/solutions/pathway-studio-biological-research
InterPro	Database of protein families, domains and functional sites in which identifiable features found in known proteins can be applied to unknown protein sequences	www.ebi.ac.uk/interpro
Enzyme nomenclature database	Repository of information related to the nomenclature of enzymes, primarily based on the recommendations of the Nomenclature Committee of the International Union of Biochemistry and Molecular Biology (IUBMB)	http://enzyme.expasy.org/
Genomics resources		
Online Mendelian Inheritance in Man (OMIM)	Database of human genes, genetic disorders and their relationships to one another	www.ncbi.nlm.nih.gov/omim
Mouse Genome Informatics (MGI)	Provides integrated access to data on the genetics, genomics and biology of the laboratory mouse	http://informatics.jax.org
Protein Data Bank (PDB)	Repository of three-dimensional structural data of proteins and nucleic acids	www.rcsb.org
Gene Expression Omnibus (GEO)	Public repository of high-throughput gene expression (such as DNA microarray) experimental data	www.ncbi.nlm.nih.gov/geo
Ensembl	Portal to vertebrate genomes (including human) with extensive cross-references to other databases	www.ensemble.org

Resource	Brief description of resource	URL
Univeral Protein Resource (UniProt)	Comprehensive catalog of information on protein sequences and functions	www.pir.uniprot.org
Gene Ontology (GO) and Gene Ontology Annotations (GOA)	GO provides a controlled vocabulary to describe gene and gene-product attributes in any organism. The GOA project provides high-quality GO annotations to proteins in UniProt	www.geneontology.org www.ebi.ac.uk/GOA

Index

Printed in the United States
By Bookmasters